DAVID ATTENBOROUGH

THE LIVING PLANET

A PORTRAIT OF THE EARTH

FONTANA/COLLINS

BRITISH BROADCASTING CORPORATION

First published in Great Britain in 1984
by William Collins Sons & Co. Ltd
and the British Broadcasting Corporation
First issued in Fontana 1985
Third impression March 1990

Set in Monophoto Ehrhardt by Jolly & Barber Ltd, Rugby
Origination by Adroit Photo Litho Ltd, Birmingham
Printed and bound by William Collins Sons & Co. Ltd, Glasgow

CONTENTS

PREFACE

This book is based on a series of programmes made for BBC Television. It and they are sequels to an earlier series and book entitled *Life on Earth*. That project attempted to describe the way in which animals and plants developed on this planet over the past three thousand million years and traced the rise of different groups of animals which culminated in the expansion of the mammals and ultimately the appearance of man.

This new book surveys the situation as it is today. It examines the manner in which the survivors of ancient groups as well as the representatives of newly-evolved ones have together colonised and adapted to the great variety of environments that exist on earth. Here and there, the two narratives may overlap slightly, but the variety of animals and plants is so immense that I have been able, in most instances, to illustrate such episodes with species other than those described in the earlier book.

I have retained the same style as before, avoiding as far as possible technical scientific terms and not encumbering the text with Latin names. However, the index has been compiled to serve as a glossary in which each organism is given not only a page reference but also its scientific name, so a reader who wishes to know precisely what family, genus or species is being referred to can discover by looking it up in the index.

The book was written at the same time as the programmes were being filmed. The one is not, therefore, the direct descendant of the other. Rather the two are cousins, both descended from the same body of research and years of travel. They therefore have the sort of differences and likenesses that you might expect from such a relationship. I hope the one may enhance the other.

PROLOGUE

The Kali Gandaki River flows through the deepest valley in the world. As you stand in Nepal beside its roaring milky waters, looking upstream towards the main range of the Himalayas, the river seems to spring from a cluster of immense snow-capped, ice-girt peaks. The tallest of them, Dhaulagiri, is over 8000 metres tall, the fifth highest mountain in the world. The summit of its immediate neighbour, Annapurna, is only 35 kilometres away and only a few metres lower. You might well suppose that the source of the river lay on the nearside, southern flanks of this immense barrier of rock and ice. Not so. The Kali Gandaki flows between the two mountains, its bed a good 6 vertical kilometres below the peaks.

The people of Nepal have, for many centuries, known that the valley is a highway that leads right through the Himalayas and up into Tibet. Every day, throughout the summer, trains of mules plod up the winding stony tracks, red horsehair plumes bobbing on their withers, red pompoms on long strings swinging from their pack saddles, carrying huge loads of barley and buckwheat, tea and cloth, up to Tibet to be traded for bales of wool and cakes of salt.

The lowest reaches of the valley are so warm and humid that the people can grow bananas. The forest has all the luxuriance of a tropical jungle. Rhinoceros munch the lush vegetation and tigers prowl through the bamboo thickets. But as you ascend into the valley proper, the plants change. By the time you reach an altitude of 1000 metres, rhododendrons have begun to appear, rather scraggy trees, some 10 metres high, with broad glossy leaves. In April, they are hung with cascades of scarlet blossom. To these magnificent flowers come tiny sunbirds, their iridescent breast feathers glinting metallically in the sun as they dip their curved beaks into the throat of the blossoms, sipping nectar and obligingly spreading pollen from tree to tree. Langur monkeys come too, but they are pillagers, grabbing handfuls of flowers and cramming them into their mouths. On the ground grow orchids and iris, trumpet-shaped arums and primulas. Where the sun penetrates the canopy and warms a boulder, you may see a little lizard, basking. And in the depths of the forest, foraging on the ground or roosting in the trees, you may catch a glimpse of one of the most glorious birds in the world, a tragopan – a

The Kali Gandaki valley, Nepal

Himalayan red panda

pheasant the size of a turkey, with ultramarine wattles and crimson feathers marvellously decorated with chains of white spots.

The luxuriance of this forest is created and sustained by abundant rain. The monsoon winds, blowing up from India, bring clouds swirling up the valley. As they get higher, they get colder and can no longer bear their loads of moisture, shedding them as torrents of rain that make the lower stretches of the Kali Gandaki one of the best-watered places on earth.

But this forest, too, has its limits. By the time you have trudged up to 2500 metres, the rhododendrons, in their turn, have disappeared except for a few patches on sheltered slopes. In their place stand conifers – Himalayan fir and Bhutan pine. Their leaves are not broad like those of the rhododendron which catch snow and sometimes break under the weight of it, but are long tough needles which shed the snow and can withstand very low temperatures. Among them, if you are very lucky, you might see a little red panda, foxy brown with a furred black-ringed tail and a grizzled head, scrambling through the branches on the lookout for birds' eggs or berries, insects or mice. It moves surefootedly over the snow-covered ground and the slippery wet branches, for the soles of its feet are covered with woolly hairs which give it a firm grip.

Another half day of walking and you emerge from the pine forest. As you leave it, you leave behind all the birds and the mammals that depended on the pine trees, directly or indirectly, for shelter and food. There is little now on the rocky hillside except a few tussocky plants and an occasional bush of buckthorn or juniper. The river itself has shrunk. Now it is a shallow stream wandering over a waste of gravel. But the valley itself is still immense, its floor still over a kilometre wide. Nor is the river much larger at any other time of the year, for there is little rain up here to feed it, most having been shed at lower altitudes. And this is the first of the puzzles of the Kali Gandaki. How could such a huge valley have been cut by such a relatively small river?

Wild animals are very scarce indeed up here. It is too cold for any lizards. Nor is there enough food to sustain langur monkeys. Indeed, you may walk all day without seeing any living creature, except for a flight of choughs or ravens and, high above, patrolling the hillsides, griffon vultures. Their presence, however, is a sure sign that other animals are here somewhere, for without them the vultures would starve. So somewhere among the rocks there must be rodents – marmots or pikas – cautiously nibbling the grass and the cushion plants that grow, here and there, on the rubbly slopes. But the grazing is so poor that it can only sustain a very small number of individual animals, and those species that do manage to survive here are all scarce. Among them are tahr, neither true sheep nor true goats but equally related to both groups. Rarer still is the animal that preys on them, the snow leopard. One of the loveliest of the cats, it has a thick creamy coat, rosetted with grey, and cushions of hair on the soles of its paws which protect them from the rough stones and from the cold. During the winter it retreats to the forests below, but during the summer it may wander as high as 5000 metres.

Though there is seldom heavy rain up here, there is an almost constant wind, bitterly

cold and sapping to the strength. You have now climbed to nearly 3000 metres and if you have come up from the lower reaches of the valley, walking every day, you will certainly be feeling the thinness of the air. It is chill in your lungs and though your chest heaves, you still feel robbed of breath. Your head may ache, you may even feel sick. A few days' rest and you will become acclimatised and the worst of these symptoms will disappear. But you will never be able to match the physical endurance of the muleteers who have come up with you and whose homes are at high altitudes.

Even the mules, at these heights, labour under their loads. The highland villagers keep a tougher stronger beast of burden, the yak. Once it roamed wild in vast herds across the Tibetan plateau. Now it is domesticated and carries loads and pulls ploughs. Its woolly coat is so thick and warm that it has to shed much of it during the summer months to prevent over-heating, and it can live permanently at higher altitudes than any other large mammal except man. Now, unexpectedly, the valley has opened out. The great peaks of Annapurna and Dhaulagiri that a few days back you glimpsed through gaps in the canopy of rhododendrons as white shining pyramids several miles above are now behind you. Ahead, the ramparts of snow are falling towards a brown streak on the horizon that is the high, dry, half-frozen plain of Tibet. You have walked clean through the greatest range of mountains in the world.

And now another extraordinary characteristic of the Kali Gandaki becomes apparent. It seems to be flowing the wrong way. Rivers, after all, normally rise in the mountains, flow down their slopes, gathering water from their tributary streams as they go, and then continue down to the plains. The Kali Gandaki does the reverse. It rises on the edge of the great plains of Tibet and heads straight for the mountains. It worms and wriggles its way downwards through the giant interlocking buttresses as the mountains on either side of it grow higher and higher. Only after it has found its way right through them does it reach a relatively flat plain and unite with the Ganges to flow down to the sea. When you stand close to its source, high on the wall of its valley, tracing its course with your eye as it writhes away like a silver snake into the distant mountains, you cannot believe that the river could have cut its way through the mountains by itself. How, then, did it ever come to follow such a course?

Clues to the answer lie at your feet, scattered among the rubble. The rock here is a crumbling, easily-split sandstone and in it lie thousand upon thousand of coiled shells. Most are only a few inches across. Some are as big as cartwheels. They are ammonites. No ammonite is alive today, but a hundred million years ago, they flourished in vast numbers. From their anatomy and the chemical constituents of the rocks in which their fossilised remains are found, we can be quite certain that they lived in the sea. Yet here, in the centre of Asia, they are not only 800 kilometres from the sea but some 4 vertical kilometres above its level.

How this came about was, until only a few decades ago, the subject of great controversy among geologists and geographers. Now, at last the broad outlines of the explanation have been deduced beyond dispute. Once, between the great continental mass of India to the south and Asia to the north, there lay a wide sea. In its waters lived the ammonites.

The upper Kali Gandaki river, Nepal

An ammonite, upper Kali Gandaki valley, Nepal

Rivers flowing from the two continents brought down layer upon layer of sediments. As the ammonites died, so their shells fell to the bottom of the sea and were covered by fresh deposits of mud and sand. But the sea was becoming narrower and narrower for, year after year and century after century, India was moving closer to Asia. As it neared, the sediments on the sea floor began to ruck and crumple so that the sea became increasingly shallow. But still the continent of India advanced. The sediments, now compacted into sandstones, limestones and mudstones, rose to form hills. Their elevation was infinitesimally slow. Nonetheless, some of the rivers that had been flowing south from Asia were unable to maintain their course over the slopes that were rising in front of them. Their waters were diverted eastwards and avoided the infant Himalayas by running round their eastern end, eventually joining the Brahmaputra. But the Kali Gandaki had enough strength to cut through the soft rocks as fast as they rose so that they formed the great cliffs of crumpled strata that can now be seen on either side of its valley.

The process continued for millions of years. Tibet, which before the collision of the continents had been a well-watered plain along the southern edge of Asia, was not only pushed upwards but gradually deprived of its rainfall by the young mountains and so changed into the high cold desert that it is today; the upper reaches of the Kali Gandaki lost much of the rain that had given the river its initial erosive power and shrank inside its vast valley; and on the site of the ancient sea there now stood the highest and newest mountains in the world containing, within their fabric, the remains of ammonites. Nor has this process stopped. India is still moving north at the rate of about 5 centimetres a year, and each year the rocky summits of the Himalayas are a millimetre higher.

This transformation of sea into land began some 65 million years ago. Though this seems inconceivably distant to us, a species that has only been in existence for less than half a million years, in terms of the history of life as a whole it was a comparatively recent event. It was, after all, some 600 million years ago that simple animals began to swim in the ancient seas; and over 200 million years since amphibians and reptiles invaded the land. Birds developed feathers and wings and took to the air a few million years afterwards and mammals evolved fur and warm blood around the same time. Sixty-five million years ago, the reptiles fell into their still-mysterious decline and mammals assumed the dominance of the land which they still hold today. So 50 million years ago, as the island continent of India approached Asia, all the major groups of animals and plants that we know today, and indeed almost all the large families within those groups, were already in existence. Each of the continents had its own multitudinous complement of inhabitants, though India, having been isolated as an immense island since just after the decline of the reptiles, was undoubtedly much poorer in advanced groups of animals than Asia. When the two eventually met and the new mountains began to rise some 40 million years ago, the animals and plants from the two old continents began to spread into the new uncolonised extension to their territory.

Jungles covered part of Asia then, as they do now, and plants and animals from them found conditions that suited them in the low foothills on the southern flanks of the new

Overleaf: *The Himalayan ranges from a satellite*

ranges. But above the foothills, there was new country at greater elevations than anywhere in Asia or India. To colonise that vacant territory organisms had to change. Sometimes the necessary adaptations were small. Langur monkeys from the warm plains were able to move up into the chilly rhododendron forests and gather leaves and fruit by doing no more than to develop slightly thicker coats to keep themselves warm. Grazing animals, such as tahr, did a similar thing. The snow leopard, coming from the same ancestral stock as the lowland leopard, not only grew a furrier coat, but became paler in colour, so making itself less conspicuous against the grey hillsides or the snow, and changed its diet from the antelope and wild cows it presumably fed on in the jungles to smaller game like tahr and marmots. Altitude posed no problems to birds like griffon vultures. They habitually soared to great heights and so had no difficulty in moving up into the huge valleys, as long as there were creatures below to provide them with meals.

The new forests and their inhabitants had long since been established by the time human beings arrived. When exactly that was we do not know, but it was certainly tens of thousands of years ago. As people moved up the valleys, they too began to respond to the new conditions. Unlike other animals, they did not have to depend solely on bodily changes to protect themselves from the cold. With the level of intelligence and the skills that are the particular possession of humanity, they were able to make warm clothing for themselves and build fires. But they could not construct a device to help them deal with the dearth of oxygen in the air. That could only be dealt with by physical changes in their bodies. And change they did. Today, their blood contains 30 per cent more corpuscles than that of people living at sea level and is in consequence able to carry more oxygen per litre. Their chests and their lungs are also particularly large, so they are able to take in more air with a single breath than a lowlander can. But even they have not yet fully adapted to the highest parts of the mountains. Above 6000 metres, women cannot bear children. The air there is so thin that they cannot get enough oxygen into their blood to sustain an infant growing in the womb.

The story of the building of the Himalayas and their subsequent colonisation by animals and plants is only one example of the many changes that are proceeding continuously all over our planet. Mountains are not only being built but simultaneously worn down by glaciers and rivers. Rivers themselves clog and change their courses. Lakes fill with sediments and become swamps and eventually plains. Nor is India the only continent to drift over the face of the globe. All have done so to some degree. As they change their positions, moving towards the equator or up to the poles, so jungle may turn to tundra and grasslands bake into desert. Each of these physical changes, in sunshine and altitude, in rainfall and temperature, demands a response from the community of plants and animals undergoing it. Some organisms will adapt and survive. Others will fail to do so and disappear.

Similar environments will call for similar adaptations and produce animals in different parts of the world which come from quite different ancestors, but which bear a marked resemblance to one another. So there are small, brilliantly coloured birds feeding from large blossoms on the slopes of the Andes which look very like the sunbirds of the

Snow leopard

Himalayas, but which belong to a quite different family of birds, and the heavy-fleeced sure-footed beast of burden that the Andean people use is the llama, a kind of camel, and not a kind of cow like the Himalayan yak.

Only two major environments have remained physically unchanged over vast periods of time – the jungle and the sea. Even here, the biological conditions have gradually altered as evolution, within or outside their frontiers, has produced new kinds of organisms and therefore presented their older inhabitants with new problems of survival.

So almost every corner of the planet, from the highest to the lowest, the warmest to the coldest, above water and below, has acquired its population of interdependent plants and animals. It is the nature of these adaptations, that have enabled living organisms to spread so widely through our varied planet, that is the theme of this book.

ONE

THE FURNACES OF THE EARTH

The titanic forces that built the Himalayas and all the other mountains on earth proceed so slowly that they are normally invisible to our eyes. But occasionally they burst into the most dramatic displays of force that the world can show. The earth begins to shake and the land explodes.

If the lava that erupts from the ground is basalt, black and heavy, then the area may have been continuously active for many centuries. Iceland is just such a place. Almost every year there is volcanic activity of some kind. Molten rock spills out from huge cracks that run right across the island. Often it is an ugly tide of hot basalt boulders that advances over the land in a creeping unstoppable flood. It creaks as the rocks cool and crack. It rattles as lumps tumble from its front edge. Sometimes the basalt is more liquid. Then a fountain of fire, orange red at the sides, piercing yellow at its centre, may spout 50 metres into the air with a sustained roar, like a gigantic jet engine. Molten basalt splashes around the vent. Lava froth is thrown high above the main plume where the howling wind catches it, cools it and blows it away to coat distant rocks with layers of grey prickly grit. If you approach upwind, much of the heat as well as the ash is blown away from you, so that you can stand within 50 metres of the vent without scorching your face, though when the wind veers, ash will begin to fall around you and large red-hot lumps land with a thud and a sizzle in the snow nearby. You must then either keep a sharp eye out for flying boulders or run for it.

Flows of cooling black lava stretch all round the vent. Walking over the corded, blistered surface, you can see in the cracks that, only a few inches beneath, it is still red hot. Here and there, gas within the lava has formed an immense bubble, the roof of which is so thin that it can easily collapse beneath your boot with a splintering crash. If, as well as such alarms, you also find yourself fighting for breath because of unseen, unsmelt poisonous gas, you will be wise to go no further. But you may now be close enough to see the most awesome sight of all – a lava river. The liquid rock surges up from the vent with such force that it forms a trembling dome. From there it gushes in a torrent, 20 metres across maybe, and streams down the slope at an astonishing speed, sometimes as much as 100 kilometres an hour. As night falls, this extraordinary scarlet

Overleaf: *Basaltic lava flow, Iceland*

river illuminates everything around it a baleful red. Its incandescent surface spurts bubbles of gas and the air above it trembles with the heat. Within a few hundred yards of its source, the edges of the flow have cooled sufficiently to solidify, so now the scarlet river runs between banks of black rock. Farther down still, the surface of the flow begins to skin over. But beneath this solid roof the lava surges on and will continue to do so for several miles more, for not only does basaltic lava remain liquid at comparatively low temperatures, but the walls and ceiling of solid rock that now surround it act as insulators, keeping in the heat. When, after days or weeks, the supply of lava from the vent stops, the river continues to flow downwards until the tunnel is drained, leaving behind it a great winding cavern. These lava tubes, as they are called, may be as high as 10 metres and run for several kilometres up the core of a lava flow.

Iceland is one of a chain of volcanic islands that runs right down the centre of the Atlantic Ocean. Northwards lies Jan Mayen; to the south, the Azores, Ascension, St Helena and Tristan da Cunha. The chain is more continuous than most maps show, for other volcanoes are erupting below the surface of the sea. All of them lie on one great ridge of volcanic rocks that runs roughly midway between Europe and Africa to the east, and the Americas to the west. Samples taken from the ocean floor on either side of the ridge show that, beneath the layers of ooze, the rock is basalt, like that erupting from the volcanoes. Basalt can be dated by chemical analysis and we now know that the farther away from the mid-ocean ridge a sample is taken, the older it is. The ridge volcanoes, in fact, are creating the ocean floor which is slowly growing away from them, on either side of the ridge.

The mechanism that produces this movement lies deep within the earth. Two hundred kilometres down, the rocks are so hot that they are plastic. Below them, the metallic core of the earth is hotter still and this causes slow, churning currents in the layers above, which rise up along the line of the ridge and then flow out on either side, dragging the basaltic ocean floor with them like solid skin on custard. Such moving segments of the earth's crust are known as plates. And most of these plates carry on them, like lumps of scum, continents.

One hundred and twenty million years ago, Africa and South America were joined together, as you might guess from the jigsaw similarity of their coastlines and as is demonstrated by the likeness of the rocks on opposite sides of the ocean. Then, about 60 million years ago, a current welling up beneath this supercontinent created a line of volcanoes. A fracture developed across the supercontinent and the two halves slowly moved apart. The line of the split is today marked by the Mid-Atlantic Ridge. Africa and South America are still moving away from one another and the Atlantic is getting wider by several centimetres each year.

Another similar ridge, extending from California southwards, was responsible for creating the floor of the eastern Pacific. A third, running from Arabia southeast towards the South Pole, produced the Indian Ocean. It was the plate on the eastern side of this ridge that dragged India away from the flank of Africa and carried it towards Asia.

The convection currents that flow up at the ridges must clearly descend again. The

Lava tube, Hawaii

lines along which they do so are where a plate meets that of a neighbouring system. It is here that continents collide. As India approached Asia, sediments on the sea floor between the two continents were crumpled and piled high to form the Himalayas, so the plate junction here is concealed beneath a mountain range. But farther along the same junction line, to the southeast, a continental mass exists on the Asian side only. The line of crustal weakness, therefore, is much more exposed and is marked by a chain of volcanoes that runs down from Sumatra through Java to New Guinea.

The descending convection current sucks down the ocean floor, creating a long deep trench. This runs along the southern coast of the Indonesian chain. As the edge of the basaltic plate descends, it takes with it water and much of the sediment that was eroded from the Indonesian land mass and had been lying on the ocean floor. This introduces a new ingredient into the melt deep in the crust, so that the lava that wells up into the Indonesian volcanoes is crucially different from the basalt issuing from a mid-ocean ridge. It is much more viscous. In consequence, it does not pour out of cracks or flow like a river, but congeals in the throat of the volcanoes. The effect is like screwing down the safety valve of a boiler.

It was one of the Indonesian volcanoes that produced the most catastrophic explosion yet recorded. In 1883, a small island named Krakatau, 7 kilometres long by 5 kilometres wide, lying in the straits between Sumatra and Java, began to emit clouds of smoke. The eruptions continued with increasing severity day after day. Ships sailing nearby had to make their way through immense rafts of pumice that floated on the surface of the sea. Ash rained down on their decks and electric flames played along their rigging. Day after day, enormous quantities of ash, pumice and lava blocks were thrown out from the crater, accompanied by deafening explosions. But the subterranean chamber from which all this material was coming was slowly emptying. At 10 a.m. on 28 August, the rock roof of the chamber, insufficiently supported by lava beneath, could bear the weight of the ocean and its floor no longer. It collapsed. Millions of tons of water fell on to the molten lava in the chamber and two-thirds of the island tumbled on top of it. The result was an explosion of such magnitude that it produced the loudest noise ever to echo round the world in recorded history. It was heard quite distinctly over 3000 kilometres away in Australia. Five thousand kilometres away, on the small island of Rodriguez, the commander of the British garrison thought it was the sound of distant gunfire and put out to sea. A tempest of wind swept away from the site and circled the earth seven times before it finally died away. Most catastrophic of all, the explosion produced an immense wave in the sea. As it travelled towards the coast of Java, it became a wall of water as high as a four-storey house. It picked up a naval gunboat, carried it bodily nearly 2 kilometres inland and dumped it on top of a hill. It overwhelmed village after village along the thickly populated coast. Over 36,000 people died.

The biggest explosion of recent years occurred on the other side of the Pacific, where the eastern edge of the Pacific plate grinds along the western coast of North America. Once again, there is continental cover on only one side of the junction, so the line of contact is not deeply buried. But because continents are made of rocks that are lighter

Anak Krakatau with Rakata behind

than basalt, they override the downwards-plunging oceanic plate and the line of volcanoes breaks through some 200 kilometres inland from the coast. And once again, the lava that rises up in them carries the sedimentary ingredient that makes them catastrophically explosive.

Until 1980, Mount St Helen's was famous for the beautiful symmetrical shape of its cone. It rose nearly 3000 metres high and was crowned with snow the year round. In March that year, warning rumbles began to come from it. A plume of steam and smoke rose from its peak, dusting its snow cap with streaks of grey. All through April, the column of smoke grew. Most ominous of all, the northern flank of the mountain, about 1000 metres below the summit, began to bulge outwards. The swelling grew at a rate of about 2 metres a day. Thousands of tons of rock were being pushed upwards and outwards. Every day there were fresh spouts of ash and smoke from the crater above. Then, at half past eight on the morning of 18 May, the mountain exploded.

The northwest face, about a cubic kilometre of it, simply blew out. The pine, fir and hemlock trees that had clothed the lower slopes of the mountain, over an area of 200 square kilometres, were laid flat, as though they were matches. An immense burgeoning black cloud rose above the mountain, towering 20 kilometres into the sky. Few people lived close to the volcano and there had been a lot of warning, but even so sixty people were killed. Geologists estimated that the force of the explosion was 2500 times as powerful as the nuclear blast that destroyed the city of Hiroshima.

Nothing can live on a volcano immediately after its eruption. If there has been an explosion, steam, smoke and poisonous gas will continue to billow up from the wreckage of rocks in the crater for weeks. Nor can any organism survive the heat of the basalt flows that issue from the volcanoes of the mid-ocean ridge. If any parts of the earth are sterile and lifeless, it must be sites such as these. But if the convection currents deep beneath the surface shift slightly, the ferocity of the volcanic furnaces begins to wane. In these later stages, a dying volcano often produces eruptions not of lava but of scalding water and steam. Part of this water existed in the magma, and part is from the natural water table of the earth's crust. It carries, dissolved in it, a great variety of chemical substances. Some will have come from the same deep source as the lava, others have been dissolved from the rocks through which the hot water passed on its way to the surface. Among them are compounds of nitrogen and of sulphur, often in such concentrations that the water can serve as food for very simple living organisms. Indeed, it is possible that the very first forms of life to appear on earth originated in just such circumstances, some 3000 million years ago.

At that unimaginably distant time, the earth had not yet acquired its oxygen-rich atmosphere and the position of the continents bore no relation to their present distribution. Volcanoes were not only very much larger than those of today, but were very much more numerous. The seas, which had condensed from clouds of steam that surrounded this new planet, were still hot and water was still gushing into them from volcanic sources deep in the crust. In these chemically rich waters, complex molecules were forming. Eventually, after an immense span of time, tiny microscopic specks of

Mount St Helen's, May 1980

living matter appeared. They had little internal structure, but they were able to convert the chemical substances in the water into their own tissues, and to reproduce themselves. These were bacteria.

Bacteria today are of many different kinds, and practise a great variety of chemical processes to maintain themselves. And they are found throughout the land, the sea and the sky. Some even still flourish in volcanic environments which may well parallel the circumstances in which they first arose.

In 1977, an American deep-sea research ship was investigating underwater volcanoes erupting from a ridge south of the Galapagos Islands. Three kilometres below the surface of the ocean they found vents on the sea floor that were spouting hot, chemically rich water into the sea. In these jets, and in the crevices of the rocks around the vents, the scientists discovered great concentrations of bacteria consuming the chemicals. The bacteria, in turn, were being fed upon by immense worms, up to 3.5 metres long and 10 centimetres in circumference. They were unlike any other worms so far encountered by science, for they had neither mouth nor gut and they fed by absorbing the bacteria through the thin skin of feathery tentacles, rich in blood vessels, that sprouted from their tip. Since these organisms live in the black depths of the ocean, they are unable to tap directly the energy of sunlight. Nor can the worms obtain it second-hand from the falling fragments of dead animals drifting down from above, since they have no mouths. Their food comes entirely from the bacteria which in turn derive their sustenance from the volcanic waters. Indeed, the worms may well be the only large animals anywhere that draw their energy entirely from volcanoes.

Alongside the worms lie huge clams 30 centimetres long which also feed on the bacteria. The rising jets of hot water create other currents which flow towards the vents across the sea floor, bringing with them organic fragments which are eaten by other organisms – strange, hitherto unknown fish and blind white crabs – clustering around the clams and the worms. So in these submarine volcanic springs, a dense and varied colony of creatures flourishes in the darkness.

Hot springs also bubble up on land. The water they produce, which originates partly from sources far below and partly from rainwater that has permeated deep into the ground, has been heated by the lava chamber and so forced up again through cracks in the rocks, like water up the spout of a boiling kettle. Sometimes, because of the particular geometry of these conduits, the upward progress is spasmodic. Water accumulates in small subterranean chambers and becomes superheated under pressure until finally it flashes into steam and a column of water spouts to the surface as a geyser. In other cases the upward flow is more regular and then the water forms a deep, perpetually brimming pool. It may be so scaldingly hot that the surface steams, but even at these temperatures bacteria flourish. Growing with them here are slightly more advanced organisms – blue-green algae. These are scarcely more complex in their internal structure than the bacteria but they do contain chlorophyll, the remarkable substance that enables them to use the energy of the sun to convert chemical substances into living tissue.

Such organisms are found in the hot springs of Yellowstone in North America.

Giant worms around submarine volcano, near the Galapagos Islands

There the algae and bacteria grow together to form slimy green or brown mats that cover the bottom of the pools.

Nothing else can survive in the hottest parts of the springs occupied by these mats, but where the pool spills over to form a stream, the water cools slightly and so allows occupation by other creatures. The algal mats here are so thick that they break the surface. This living dam diverts the main flow to another freer part. As the water slowly trickles through, it cools further and above it assemble clouds of brine flies. If parts of the algae are cooler than 40°C, the flies settle and begin to graze voraciously. Some of them mate and lay their eggs on the algae and soon there are grubs feeding alongside their parents. But they are working towards their own destruction or that of their descendants, for as they chew away at the mat, so they weaken it. Eventually, it breaks up, the channel clears and much hotter water from the pool gushes down it, sweeping away the remains of the algae and killing all the grubs that were feeding on it. But enough will have hatched for the flies to survive this setback and to start the process all over again in another part of the spring.

In colder parts of the world, the dwindling heat of a volcano may represent not a hazard but a haven. The line of volcanoes that built the Andes along the junction of the South American and eastern Pacific plates continues south and east into the southern ocean to form several small arcs of volcanic islands. Bellinghausen is one of a group called the South Sandwich Islands. The ferocious Antarctic seas have cut into its base, creating, on one side, a cliff which displays, with textbook clarity, alternating layers of ash and lava, cut through with zig-zag lines of lava-filled pipes. Ice floes rim it like a tattered white skirt and sheets of snow drape its slopes. Battalions of Adelie penguins march all over this white parade ground. If you climb up through their ranks to the top of the volcano, you find a vast gaping pit, half a kilometre across. Its floor is filled with snow, icicles hang from the jutting rocks in its throat and snow petrels, elegant pure-white birds, nest in the crags just beneath the crater's lip. But its volcanic fires have not been totally extinguished. In one or two places around the rim, steam and gas still spurt from cracks, filling the air with the stench of hydrogen sulphide and coating the boulders with brilliant yellow encrustations of sulphur. The ground around the vent is warm to the touch so, as the polar gales bite into you, it is a pleasant place to crouch, in spite of the smell. And on the rocks at your feet, surrounded by snow, there are lush cushions of mosses and liverworts.

These few small patches are the only places in the entire island where it is warm enough for plants to grow. The islands are as isolated as any in the world. The Antarctic continent and the tip of South America are both some 2000 kilometres away. Yet the spores of these simple plants are so widely dispersed by winds throughout the atmosphere of the world that even these tiny isolated sites in this hostile island are colonised just as soon as they become habitable.

It is not only in the bitterly cold parts of the world that organisms take advantage of volcanic heat. Even tropical creatures have learned how to exploit it. The megapodes are a group of birds living from Indonesia to the western Pacific which have developed

Maleo birds on volcanic sands, Sulawesi

extremely ingenious methods of incubating their eggs. Typical of them is the mallee fowl of Australia. When this remarkable bird nests, it first digs an enormous pit that may be 4 metres across, fills it with decaying leaves and then piles sand on top of it. Into this great heap, the female excavates a tunnel and there she lays her eggs. The male fills the tunnel with sand and relies on the heat produced by the rotting vegetation to keep the eggs warm. But he does not abandon them. On the contrary. Several times a day, he returns to the mound and pokes his beak into the sand. His tongue is so sensitive that he can detect a change in heat of one-tenth of a degree. If he considers that the sand is too cool for the eggs, he will pile on more; if too hot, he will scrape it away. Eventually, after an unusually long incubation, the young mallee fowl chicks dig their way up to the surface of the mound, emerge fully-feathered and scamper away.

The mallee fowl, however, has a relative in the Indonesian island of Sulawesi called the maleo. This creature buries its eggs in black volcanic sand at the head of beaches. Being black, this sand absorbs the heat and gets quite hot enough in the sunshine to incubate the eggs. Other maleo have left the coast and colonised the slopes of a volcano inland. There they have discovered large areas of ground that are permanently heated by volcanic steam; and there a whole colony regularly lays its eggs. A dying volcano has become an artificial incubator.

Eventually, as the plates of the earth's crust move and the currents beneath shift, volcanoes do become completely extinct. The ground cools and animals and plants from the surrounding countryside move in to colonise the new, sterile rocks and the devastated land. Basalt flows present considerable problems to the colonists. Their shiny blistered surface is so smooth that water runs off it and there are few crevices into which seedlings can insinuate their young roots. Some flows may remain totally bare for centuries. The species of flowering plant that makes the first pioneering invasion differs from one part of the world to another. In the Galapagos, where the flora is derived primarily from South America, it is often a cactus which gains the first root-hold. Specially adapted to conserve every particle of moisture and living normally in deserts, it manages to survive the roasting temperatures out on the black lava. In Hawaii, the pioneer is a less obvious conserver of water, the ohia lehua tree. Its roots manage to penetrate deep into the lava flow to gather moisture. Often, they find a way down into the empty cavern, the lava tube, that runs down the centre of most of these flows. Down there, the roots hang from the ceiling like huge brown bell-ropes. Rain water, running from the lava surface, trickles down the cracks and over the roots and drips on to the floor. Away from the evaporating rays of the sun, it lingers in pools making the air in the cave dank and humid.

A lava tube is an eerie place to explore. Since neither rain storms nor frost can reach it, nothing erodes the surface of its walls or floor. It looks exactly as it must have done when the last trickle of lava was draining away and its floor was still hot enough to incinerate anything that touched it. Congealed drips of lava hang from the ceiling like stalactites. The floor is covered by a stream of lava like solidified porridge. In some places, where it swept over a barrier of some kind, it has left behind a solidified cascade.

Cactus on lava, Galapagos

As a sudden surge swept through it, the lava river rose temporarily, cooled particularly quickly and so left a smooth tide mark along the wall.

Several kinds of creatures have taken up permanent residence in these strange places. In the tiny hairs that cover the hanging roots, and feeding on them, are several kinds of insects including crickets, springtails and beetles. And preying on them are spiders. But these creatures are not exactly the same as their close relatives that live in the open air elsewhere on the island. Many of them have lost their eyes and wings. It seems that once a part of an animal's anatomy loses its function, its development is a waste of bodily energy. Individuals that do not squander their resources in this way, therefore, have an advantage over those that continue to do so. So, over generations, there is a tendency for useless organs to be reduced in size and finally to disappear. On the other hand, it is a positive advantage in the blackness of the cave to have long antennae and legs so that a creature can detect obstacles or food around it. And these lava-tube creatures do indeed have unusually long legs and antennae.

The wastelands produced by continental eruptions seem to be easier to colonise than smooth basaltic flows, for it is not difficult for plants to get a root-hold on ash or rubbly lava. The great desert that was created when Mount St Helen's blew off its side is already being reclaimed by plants. In the corners of the mud banks and beneath boulders, you can find small accumulations of fluffy airborne seeds. Many belong to willow-herb, a waist-high plant which produces a spike of handsome purple flowers. Its seeds are so light and fluffy that they float on the wind for hundreds of miles. In Europe, during the last World War, willow-herb appeared on bombed sites within weeks, cloaking the broken masonry with colour. In North America, the plant is known as fire-weed, for it is one of the first to appear among the blackened stumps in the wake of a forest fire. And it is an equally enterprising colonist of the sites devastated by a volcanic eruption.

Even so, it may be several years before it succeeds in covering the naked slopes of Mount St Helen's. This is not so much because the volcanic ash lacks sustenance, but because the muds and gravels are so loosely compacted that a rainstorm or a strong wind quickly shifts the surface and uproots any seedlings. But even though there are few plants growing on it yet, there are animals to be found. The same winds that transport the willow-herb seeds also carry up moths, flies and even dragonflies. These strays, transported here by accident, are doomed to an early death, for there is virtually nothing for them to eat except one another. But they will, nevertheless, provide a basis for more permanent colonisation. When they die, the fragments of their bodies are blown with the seeds into crevices and corners. There they decay and the nutrients from their tiny corpses are absorbed into the ash beneath them, so that when the seeds germinate, they find a nutritious element immediately beneath them in the otherwise sterile, unweathered volcanic dust.

Krakatau shows how complete a recovery can be. Fifty years after the catastrophe, a small vent spouting fire arose from the sea. The people called it Anak – the child – of Krakatau. Already it has thickets of casuarina and wild sugar cane growing on its flanks. A remnant of the old island, now called Rakata, lies a mile or so away across the

Fire-weed, Mount St Helen's

sea. The slopes that a century ago were bare are now covered by a dense tropical forest. Some of the seeds from which it sprang must have floated here across the sea. Others were carried by the wind or brought on the feet or in the stomachs of birds. In this forest live many winged creatures – birds, butterflies and other insects – that clearly had little difficulty in reaching the island from the mainland a mere 40 kilometres away. Pythons, monitor lizards and rats have also arrived here, perhaps on floating rafts of vegetation that frequently get swept down tropical rivers. But evidence of the newness of the forest, and the cataclysm that preceded it, is easy to find. The tree roots cover the surface of the ground with a lattice that clasps the earth together, but here and there, a stream has undermined them, and a tree has toppled to reveal the still loose and powdery volcanic dust beneath. Once the plant cover has been broken in this way, the loose ash is easily eroded by the stream and a narrow gorge, 6 or 7 metres deep, appears beneath a roof of interlaced roots. But these breaks are the exception. The tropical forest has, within a century, reclaimed Krakatau. Without much doubt, the coniferous forest, in another century, will have reclaimed Mount St Helen's.

So the wounds inflicted on the land by volcanoes eventually heal. Although volcanoes may seem, on the short scale by which man experiences time, the most terrifyingly destructive aspect of the natural world, in the longer view they are the great creators. They have constructed new islands, like Iceland, Hawaii and the Galapagos, and built mountains like Mount St Helen's and the Andes. And it is the great shifts in the continents of the earth, with which they are associated, that set in train the long sequence of environmental changes and, over millennia, provide animals and plants with new opportunities to build their communities.

TWO

THE FROZEN WORLD

Nothing can live permanently on the summits of the high Himalayas, or on any of the other great mountain peaks of the world. They are scoured ceaselessly by the most ferocious winds on earth which at times reach velocities of over 300 kilometres an hour; and they are blighted by a cold of lethal intensity.

It may seem paradoxical that the parts of the earth that are closest to the sun are among the coldest. Warmth in the air, however, is created when the sun's rays, shining through it, give additional energy to the atomic particles of atmospheric gases, causing them to collide more frequently with one another. Each tiny collision gives off heat. The thinner the air, the more widely spread those atoms are and the more infrequent the collisions and so, in consequence, the colder the air remains.

And cold kills. If it permeates the body of a plant or an animal so thoroughly that the liquid in its cells freezes, then, with very few exceptions, cell walls will rupture, just as frozen domestic pipes will burst, and its tissues are physically destroyed. But cold can kill animals long before it freezes them solid. Most animals, including insects, amphibians and reptiles, draw their heat directly from their surroundings. They are, therefore, sometimes called 'cold-blooded', but the term is a misleading one for their blood is often far from cold. Many lizards, for example, organise their sun-bathing so effectively that they keep their bodies warmer than that of a man throughout the day, even though they cool down considerably during the night. Such creatures can tolerate a considerable drop in body temperature, but even they will die long before they are chilled to freezing point. As their temperature falls, the chemical processes which produce the energy in their bodies slow down, so that the animals become more and more sluggish. Eventually, at about 4°C above freezing point, their nerve membranes lose the semi-liquid character necessary to transmit their minute electric signals, the animal loses its bodily co-ordination and it dies.

Birds and mammals have a better chance of survival in the cold, for they generate their own heat internally. But they pay a very high price for it. A man even on a reasonably warm day uses half his intake of food to keep his body warm. In really cold circumstances with inadequate clothing, he cannot replace the heat at the rate he loses it, no matter how much he eats. His brain and the other highly complex organs of his body cannot tolerate more than a few degrees variation in temperature, and if his body cools to a level that would make reptiles merely lethargic, he dies.

So the great mountain peaks, where the temperature may fall to minus 20°C, are without life – except for small creatures that may be accidentally blown up there, and the occasional human being who, perhaps even more inexplicably, has decided to venture up there of his own accord.

A climber, descending from such a peak, is unlikely to see any other living thing among the cliffs of ice and frozen rock until he is a thousand metres or so below the summit. The first organism he can find, maybe even as high as 7000 metres, will almost certainly be a thin blister-like skin on a rock – a lichen. This is not a single species of plant but two very different ones living in the closest possible intimacy. One is an alga, the other a fungus. The fungus produces acids which etch the surface of the rock, enabling the colony to get some purchase on the smooth surface and dissolving the minerals into a chemical form that the alga can absorb. The fungus also provides a spongy framework for the colony which absorbs moisture from the air. The alga, with the help of sunshine, synthesises the rock minerals, the water and carbon dioxide from the air into food substances on which both it and the fungus feed. Both plants reproduce separately and the next generations have to re-establish the liaison afresh. The partnership, however, is not an equal one. Sometimes the fungal threads inside the lichen wrap themselves around the algal cells and consume them; and whereas the alga if separated from the fungus can lead an independent life, the fungus cannot survive without the alga. The fungus seems to be using the alga as a slave to enable it to colonise these bleak areas otherwise closed to it. Many species of alga and fungus form these alliances, but particular pairings are so habitual that the resulting organisms are regarded as regular species with their own characteristic shape, colour and rock preferences.

There are some 16,000 species of lichens in the world. All are slow-growing, but those that encrust the rocks of mountain peaks are particularly so. At high altitudes, there may be only a single day in a whole year when growth is possible and a lichen may take as long as sixty years to cover just one square centimetre. Lichens as big as plates, which are very common, are therefore likely to be hundreds if not thousands of years old.

The snowfields that drape the upper flanks of mountains may appear to be even more devoid of life than the rocks around them. Yet all are not a pristine white. In the Himalayas and the Andes, in the Alps and the mountains of the Antarctic, some stretches are as pink as a slice of water melon. The sight is difficult to believe. To climb in such a place, you need snow goggles to protect your eyes and you may well think that the strangely coloured blotches and smears in the snow slopes around you are shadows, or some trick of your dazzled eyes. If you examine a handful of this remarkable snow with your naked eye, there is nothing unusual to be seen – except its undoubted pinkness. Only with a microscope can you discover, among the frozen particles, the cause of the colour – a great number of tiny single-celled organisms. These, too, are algae. Each

Red snow, Antarctica

contains green particles with which it photosynthesises, but this colour is masked by a pervasive red pigment which may well serve the alga in the same way as your snow goggles serve you – by filtering off the harmful ultra-violet rays in the sunshine.

At one stage in its life, each of these algal cells has a tiny beating thread, a flagellum, which enables it to move through the snow to reach a level, just below the surface, where there is exactly the amount of light that best suits it. There, sheltered from the wind by the snow itself, temperatures are not as cripplingly low as they are in the open air. Even so, the snow algae need protection from the cold and they contain a chemical substance that remains liquid at temperatures several degrees below the freezing point of water.

These tiny plants take nothing from the world except sunlight and a minute quantity of nutrients that are dissolved in the snow. They feed on no other living thing and nothing feeds on them. They scarcely modify their surroundings except to bring a blush to the snow. They simply exist, testifying to the moving fact that life even at its simplest level occurs, apparently, just for its own sake.

Other creatures of a more complex kind also inhabit the snowfields, among them tiny worms and primitive insects such as bristletails, springtails and grylloblattids. Often they flourish in such numbers that they too stain the snow, but black and not pink. This dark pigmentation may be of positive value to them, for dark colours absorb heat whereas light ones reflect it. Even with this aid to warmth, however, they have to live for most of the time with their bodies close to freezing point. They, too, contain anti-freeze and their physiological processes are so adapted to low-temperature operation that if they are suddenly warmed, as they will be if you put one on your hand, they cannot function properly and they die. They are unable to synthesise food as snow algae can, and feed instead on pollen grains and the bodies of dead insects that happen to be carried up by the winds from the valleys below.

As you might expect from such frigid organisms, their whole life is conducted at an extremely slow pace. An egg from a grylloblattid takes a year to hatch and the larva needs five years to grow to adulthood. All of them are wingless. This is scarcely surprising for insect wings, to be effective, have to be beaten very rapidly and no insect muscle is able to do that at a low temperature. It simply cannot generate enough energy. One of these strange creatures, a wingless scorpion fly, has evolved its own method of compensating for its loss of flight; one that requires no swift muscular reaction. It has a tiny elastic pad in the joints of its legs which its muscles slowly compress and then lock in position. If the insect is threatened by an enemy, it suddenly releases the pad which expands explosively, so that it makes a great soaring jump.

Among the boulders beside the snowfields huddle small, cushion-like plants – mountain pinks, saxifrages, gentians and mosses. They hug the steep ground to keep out of the wind, but their roots are long and sometimes reach down almost a metre into the ground so that the plant can resist the tug of the gales and maintain its position among the shifting stones. Stems and leaves are packed tightly together in the cushion, giving one another support and protection against the cold. Some plants are even able

Rock hyrax, Mount Kenya

to use their food reserves to generate a little heat and can melt the snow around them. All grow extremely slowly. One or two tiny leaves may be as much as a plant can produce in a whole year and it may take a decade to garner enough reserves to be able to sprout flowers.

Still farther down the mountainside, where the cold is a little less intense and ridges running down from the summits provide a little more shelter from the wind, where the ground is less steep so that the boulders and splintered rocks quarried by the frost from the mountain faces lie with some stability, it is at last possible for plants to gain an adequate roothold and to raise their stems more than an inch or so above the ground. There, in particularly favourable patches, they create a near-continuous cover of green. Even at these comparatively low levels, however, protection against the cold is essential.

In the high valleys on the flanks of Mount Kenya in Africa grow some of the most spectacular of all mountain plants. They are giants. Some are groundsels, others lobelias. The groundsels stand over 6 metres high and look like enormous cabbages on trunks. Their leaves, when they die, remain attached to the main stem, forming a thick air-trapping muff which fends off much of the cold. One of the lobelias grows into a tall pillar 8 metres high on which small blue flowers appear, interspersed with hairy grey leaves so long and thin that they give the column a furry appearance. While they do not completely trap the air, they prevent its free circulation around the column and give it real protection against the frost at night. Another lobelia grows close to the ground in a huge rosette, half a metre across. Its centre is filled with water. As evening falls, this freezes over. The plate of ice wards off further chilling of the water beneath so that the plant has, in effect, a liquid protective jacket around its central bud. When the sun rises in the morning, it melts the ice cover. Now the lobelia has a different problem. Because it is growing close to the equator and at such an altitude that the atmosphere is very thin, the sun's rays are extremely strong. There is, therefore, a real risk that the water in the lobelia's central chalice may evaporate and leave it unprotected. The water, however, is not simply rainwater that happens to have accumulated there. It has been secreted by the plant itself and is slightly slimy, for it contains pectin, a gelatinous substance which greatly reduces evaporation. So the plant retains its liquid insulation even on the hottest days in readiness for the coldest nights.

The great size of these African lobelias and groundsels is in remarkable contrast to the dwarf forms that live higher up on the mountain and also to the lobelias and groundsels that grow elsewhere in the world and are nearly all tiny. In the Andes, some members of the pineapple family have developed into giants in a similar way. Both sites are high and close to the equator so these two factors may combine to make such huge dimensions advantageous, but so far, botanists have been unable to understand exactly why this should be so.

Green leaves of any kind growing sparsely on a mountain side will, very soon, tempt animals to come up and nibble them. Such adventurers must themselves take precautions against the cold. On Mount Kenya, hyrax, the size of rabbits but related to elephants, munch the lobelia leaves. They have much longer hair than their lowland relatives.

Lobelia, Mount Kenya

Their equivalents in the Andes, chinchillas, are about the same size with similar shape, habits and diet, but are rodents and not at all closely related. They have developed one of the densest and silkiest furs produced by any animal anywhere. Another Andean creature, a wild camel called the vicuna, produces one of the most valued of all wools. Its thick fine fleece insulates it so well that it is in danger of overheating, if it is particularly energetic. Accordingly, its fleece does not cover it completely. Small patches on the inside of its thighs and in its groin are almost naked. If the animal is too hot it stands in a position that exposes these areas to the air, so that they cool rapidly. If it is cold, then it assumes a posture in which the pairs of bare patches on thigh and groin are pressed together and, in effect, its covering of wool is unbroken.

Thick coats of fur or wool, however, are not the only way to conserve heat. The proportions of a body can also have a considerable effect in the matter. Long thin extremities are easily chilled so mountain animals tend to have small ears and shorter rather than longer limbs. The most heat-retentive shape of all is a sphere, and the more globular an animal becomes, the better it will hold its warmth. Size is also of consequence. Heat is lost by radiation from the surface of the body. The smaller that surface is compared with the volume of a body, the better that body will retain its warmth. So a large sphere will remain warmer longer than a small one. The effect of this is that individual animals of a particular species living in a cold climate tend to be bigger than others of the same species living in warmer areas. The puma, for example, is found throughout the Americas, from Alaska in the north, through the Rockies and the Andes down to the jungles of Amazonia. The lowland dwellers are pygmies compared with those in the mountains.

If you want to find vicuna and chinchilla in the central Andes, around the equator, you will have to climb up to the snowline, around 5000 metres above sea level. But if you travel south along the Andean chain, the snowline becomes lower and lower. By the time you reach Patagonia and the southern tip of the continent you will find permanent snow at an altitude of only a few hundred metres, and glaciers that flow directly into the sea.

The reason for this is not complicated. At the equator the sun's rays strike the earth four-square. But further round the curve of the globe towards the poles, the rays become more and more glancing. So the amount of sunshine striking a square metre of flat land on the equator is spread over a much greater area further south. The rays themselves are also less warming near the poles for, since they strike the earth's atmosphere at an angle, their path through it is much longer and they have lost more of their energy by the time they reach the earth. So the sea coasts of Antarctica are as cold and as desolate as the high summits of the equatorial Andes.

Creatures that live in the Antarctic have to face not only extreme cold but also prolonged darkness. Because the axis of the spinning earth is slightly tilted with respect to the sun, the polar regions go through major seasonal changes as the earth makes its annual orbit around the sun. When summer starts, the days become lighter for longer until by midsummer the sun is visible continuously throughout twenty-four hours.

Vicunas, Peru

The price for this boon, however, is that at the end of summer they shorten and eventually in midwinter the land is continuously dark for weeks on end.

Lichens, once again, are among the few organisms that can tolerate such oppressive conditions, and they do so spectacularly well. Over 400 species of them grow on the rocks of the Antarctic. Some are flat skins, others crusts and curling strips. The commonest is black and so, like snow insects, absorbs the maximum heat from the meagre light. Many form curling tangles of bristly or slightly rubbery branching hairs. These miniature forests contain their own communities of tiny animals. Springtails and flocks of mites, each no bigger than a pinhead, clamber slowly through the branches, browsing. Other carnivorous mites plod after them in a slightly more lively way, seizing them with their jaws and carrying them away to eat them alive. There are a few species of moss, some of which can endure being frozen solid for weeks on end; an alga that, almost unbelievably, manages to penetrate cleavage cracks in some of the rocks and lives inside using the light that filters down to it through the translucent minerals; and just two species of flowering plants – a stunted grass, and a kind of carnation. None of these plants grow in sufficient profusion to provide food for animals of any size. The creatures that live on the shores and ice-fields of Antarctica must derive their sustenance, directly or indirectly, from plants that grow not on land but in the sea.

The waters of the southern ocean are warmer than the land for they are constantly in circulation, moving back and forth between the Antarctic and the more temperate regions farther north. Being salty, they do not turn to ice until their temperature falls to a degree or so below 0°C. Cold water, however, contains more dissolved oxygen than warm water does and the Antarctic seas, in consequence, are rich in floating algae. These are eaten by immense numbers of the swimming shrimps that are known as krill, and they, in turn, together with small fish, are the food of the larger animals of Antarctica – seal, fur seals and penguins. To collect this food, however, these creatures must go to sea and there they have to have quite different protection against the cold from that used by land animals. Water absorbs more heat and conducts it very much more efficiently than air, so a swimmer is chilled far sooner than a walker. Furthermore, air trapped in fur has its limitations as an insulator in water.

The fur seal, which is a kind of sea-lion and not a true seal, has retained much of the hair of its land-living four-footed ancestors. It is so dense and warm out of water that it was eagerly sought by human beings for fur coats. The thick underlayer, which gives it its particular softness, is so extremely fine that it retains air even when the animal is in water. But were the fur seal to dive to any depth, the pressure of the water would compress the air so much that the animal would be virtually uninsulated. So fur seals do not often dive deeply in search of their food.

True seals, however, are better equipped to deal with the cold. Their fur is very sparse. It protects their skin from abrasions and it also retains a more or less permanent layer of water when they are swimming which, like a full swimming costume, reduces their heat loss to some degree. But they are also insulated with blubber, a thick layer of oily fat just beneath their skin. A fur seal has such layers in patches on its body which

Adelie penguins, Antarctica

serve it as food reserves. Seals, however, have developed their blubber into a continuous blanket that wraps their entire body and it remains efficient, no matter how deep they swim.

The Weddell seal regularly makes dives lasting up to a quarter of an hour to depths of 300 metres or more. There, in the blackness, it pursues fish using sonar – emitting high-pitched squeaks and detecting the position of fish by the echoes they cause. It is the most southerly living of all mammals and is not deterred by the ice which covers the seas around the continent in the winter. It breathes either from pockets of air trapped beneath the ice or by maintaining small holes in the ice floes, keeping them open by chewing the edges. The crab-eater seal, the most numerous of all the family in the Antarctic, feeds entirely on krill, and has special cusps on its cheek teeth that act as sieves, keeping the krill in the mouth while the unwanted water is expelled. The leopard seal, growing to a length of 3.5 metres, is slim and sinuous and eats meat of all kinds – fish, krill, young seals of other species and, occasionally, penguins.

The biggest of all is the elephant seal. This is a truly monstrous creature which may weigh as much as 4000 kilos. When an angry male rears up at you on a beach, it towers to over double your height. It owes its name not only to its immensity but also to an inflatable trunk on the top of its nose which it can blow up into a huge bladder. The elephant seal also dives to great depths and there feeds on squid. It has the thickest blubber of all. Every year it moults the thin layer of hair that covers its skin. Growing new hair requires an abundant supply of blood close to the surface of the skin and vessels therefore open up through the fat. With their blubber-wrap punctured in this way and blood circulating close to the surface of their body, the animals are no longer efficiently insulated so they have to come out of the water. A few months earlier, at breeding time, the males had fought one another ferociously on the beaches. Now they suppress their antagonisms and pile up on top of one another in the mud wallows in order to keep warm, while their skin peels off in untidy tatters.

The birds of the Antarctic, like all birds, are well protected against the cold, for feathers, in air, are the finest insulation of all. But most birds do not have feathers on their legs and the gulls that perch so nonchalantly on icebergs seem to be risking the leakage of their precious body warmth through their naked shanks and toes. The artery that carries blood down their legs, however, does not run directly to their toes. Instead, it develops, some way down the leg, a network of capillaries which wraps round the vein carrying blood back to the body from the lower foot. Heat from the arterial blood, before it is lost to the outside world, is transferred to the cold venous blood and so sent back to the body and conserved. The arterial blood itself, now cool, continues down to the feet. So the legs operate, in effect, as independent low-temperature units and their relatively simple movements are carried out by physiological processes that are adapted to work in the cold.

The characteristic birds of the Antarctic, which are often taken, indeed, as the very symbol of the far frozen south, are, of course, the penguins. In fact, the evidence of fossils suggests that though the family originated in the southern hemisphere, it did so

Elephant seal, South Georgia

in the warmer parts of it. Even today, some species of penguin live in the relatively warm waters of southern Africa and south Australia. One lives actually on the equator, in the Galapagos. Penguins are superbly adapted to the swimming life. Their wings have become modified into flippers with which they beat the water and drive themselves along. Their feet are used for steering and are placed in the best position for the purpose, at the very end of their body. This gives them their characteristic upright stance when they come out of water. Swimming everywhere demands good insulation and the penguins have developed their feathers to provide it. They are very long and thin, with tips that turn downwards towards the body. The shaft not only has filaments along the blade but, at the base, fluffy tufts that mat together and form a layer that is virtually impenetrable to wind or water. This feather coat covers more of their body than does that of any other bird. It extends low down on the legs of most of them, and the little Adelie penguin, which is one of only two species that lives on Antarctica, even has feathers growing on its stubby beak. Underneath this feather coat is a layer of blubber. So effectively protected are penguins that, like the vicuna, they run a real risk of overheating. They deal with that when necessary by ruffling their feathers and by holding their flippers out from their body to increase their radiating surface.

With such efficient insulation penguins have been able to colonise most of the waters of the southern oceans and in places they flourish in astronomic numbers. On Zavodovski, a small volcanic island in the South Sandwich group only 6 kilometres across, 14 million pairs of chinstrap penguins nest. They are small creatures, standing no higher than a man's knees. At the beginning of the Antarctic summer they come in to land, the huge swell hurling them on to the rocks with such violence that they seem certain to be smashed. But they have the resilience of rubber balls and as the surf drains back from the rocks it leaves them unharmed and undismayed and they waddle perkily inland. There on the bare volcanic ash they excavate simple scoops, squabbling ferociously and with ear-splitting shrieks over the pebbles with which they want to line them. In these meagre scrapes, they lay two eggs. The male incubates them while the female goes down to feed. If, as sometimes happens, the pair have chosen to nest in a gully where the ash is underlain by ice, then the heat of his body will melt the ice which drains away leaving him with his eggs sitting, in a rather bewildered way, in a deep hole. When the young hatch, the parents take it in turn to feed them. They grow rapidly so that by the time the short Antarctic summer is over, they are fully fledged and capable of swimming and feeding for themselves.

The largest of all the penguins is the emperor. It stands waist-high to a man and weighs 16 kilos, which makes it one of the biggest and heaviest of all sea birds. This great size may well be an adaptation to the cold, for the emperor lives and breeds on the Antarctic continent itself and is the only animal of any kind that is able to live through the extreme cold of the Antarctic interior during the winter. However, while their size undoubtedly helps them to retain heat, it also causes them great difficulties. Penguin chicks cannot feed themselves until they are fully developed and have their seagoing feathers. But large chicks take a long time to hatch and grow to their full size. Emperor

Emperor penguins with chick, Antarctica

penguin chicks cannot achieve this within the few weeks of the Antarctic summer as chinstrap or other smaller penguins manage to do. The emperors have dealt with this difficulty by adopting a breeding timetable that is exactly the reverse of that followed by most other birds. Instead of laying in the spring and rearing their offspring through the warmer months of summer when food is easy to get, the emperors start the whole process at the beginning of winter.

They spend the summer feeding at sea and at the end of it are as fat and as fit as they will ever be. In March, a few weeks before the long darkness of the winter begins, the adults come ashore on the sea-ice. It already extends a considerable way out from the shore and the penguins have to walk south for many miles to reach their traditional breeding grounds close to the coast. Throughout the dark months of April and May, the birds display to one another and finally mate. The pair claim no particular territory for themselves, nor make any nest, for they are standing on sea-ice and there is no vegetation or stones with which to line a scrape. The female produces just one egg, large and very rich in yolk. As soon as it emerges, she must lift it from the surface of the ice before it freezes. She does this by pushing it towards her toes with the underside of her beak and taking it up on to the top of her feet. There it is covered by a fold of feathered skin that hangs down from her abdomen. Almost immediately her mate comes to her and in a ceremony that is the climax of the breeding ritual, takes the egg from her on to his own feet and tucks it beneath his own apron. Her immediate task is done. She leaves him and sets off through the deepening darkness to the edge of the sea-ice where she can at last feed. But winter is now more advanced and the ice extends even farther away from the coast. She may, therefore, have to travel as much as 150 kilometres before she reaches open water.

Meanwhile, her mate has remained standing upright, his precious egg on his feet, warm beneath his stomach fold. He does little, shuffling around to huddle together with the rest of the incubating males so that they give one another a little protection, turning his back against the driving snow and the screaming winds. He has no energy to waste on unnecessary movement or needless displays. When he first arrived here from the sea, he had a thick layer of fat beneath his feathers that made up almost half his body weight. He has already drawn upon that to sustain him through the exertions of his courtship. Now it must last him for another two months while he incubates his egg.

At last, sixty days after it was laid, the egg hatches. The young chick is not yet able to generate its own body heat and remains squatting on its father's feet, beneath his apron and warmed by his body. Almost unbelievably, the male manages to find from his stomach enough food to regurgitate and provide a meal to his newly-hatched offspring. And then, with extraordinary accuracy of timing, the female reappears. She has put on a great deal of weight. There is no nest site for her to remember. The male, in any case, may have shuffled quite a long way across the ice from where she last left him. She finds him by calling and recognising the individual tones of his reply. As soon as the pair are reunited, the female gives their chick a feed of regurgitated half-digested fish. The reunion is a critical one. If she had been caught by a leopard seal and failed to return, the chick would die of starvation within the next few days. Even if she is a day or so late, she may not be in time to provide it with the food it urgently needs. It will have perished before she reaches it.

The male, having stood and starved for weeks, is now free to find food for himself. Leaving the chick in the charge of his mate, he sets off for the sea. He is pitifully thin, having lost at least a third of his weight, but if he succeeds in reaching the edge of the ice, he dives into the sea and begins to gorge. For two weeks, he has a holiday. Then with a stomach and a crop full of fish, he sets off on the long trek back to his chick.

The youngster has had nothing more to eat than the fish carried in by the female, and some juice from her stomach. It is more than ready for further food from its father. It is still dressed in its chick's coat of fluffy grey feathers. All the chicks stand together in a huddle, but each is nonetheless recognisable individually to its parents by its voice. For the remaining weeks of the winter, the parents take it in turn to go fishing and bring back food for their youngster. At long last, the horizon begins to lighten, the temperature rises infinitesimally and cracks begin to appear in the sea-ice. Leads of open water develop closer and closer to the nurseries. Eventually one comes close enough for the chicks to reach. They shuffle down to it and dive in, excellent swimmers from the moment they hit the water. The adult birds join them in the feasting. They have a mere two months to restore their fat reserves before they must start the whole cycle over again.

The breeding process has been fraught with dangers and difficulties. Safety margins have been tiny. Weather just fractionally worse than usual, fishing just a little less productive, a parent just a day or so late – any such variation could result in the death of the chick. The majority, in fact, do die. If four out of ten of them reach maturity, the emperors have had a good year.

Antarctica has not always been such a desolate land. Its rocks contain the fossil remains of ferns and forest trees, small primitive mammals and dinosaurs. They were all flourishing at a time, over 140 million years ago, when this land, together with South America, Australia and New Zealand, was part of a great southern supercontinent which lay much closer to the equator with a much warmer climate. But when the moving oceanic plates began to split the supercontinent apart, Antarctica, attached at first to Australia, drifted south. At this time the southern polar regions were covered by sea. The waters must have been cold because of the slanting angle at which the sun's rays struck them, but their circulation to warmer parts of the globe probably prevented them from freezing. But as Antarctica, separated now from Australia, continued south and eventually came to lie over the Pole itself, the situation changed. The land must have rapidly become far too cold for the dinosaurs and other land animals, even had they survived as long as this, for it could not, of course, be rewarmed as the circulating seas had been. When snow fell on the continent in winter, it remained and that, in itself, chilled the land still further, for the whiteness reflected 90 per cent of the heat in the already feeble rays of the sun. So the snow accumulated, year after year, and under the pressure of its own weight, turned to ice.

Today, ice covers the entire continent except for the tips of a few mountains that project through it and one or two strips of land near the coast. In places, it is 4.5 kilometres thick. It covers an area as big as the whole of western Europe and rises in a great dome which, at its highest point, is 4000 metres above the level of the sea. It contains 90 per cent of the world's entire supply of fresh water. Were it to melt, the level of the seas around the world would rise by 55 metres.

While Antarctica was drifting south, the continents in the northern hemisphere were changing their positions too. The North Pole, at that remote period, was also covered by freely circulating water, but Eurasia, North America and Greenland moved towards it, forming a tightening ring. This may also have interrupted the free flow of currents and interfered with the rewarming of the waters. This time, the seas themselves froze over and still today the North Pole is covered, not by a continent, but by sea-ice.

The cooling effect produced by these changes in the position of the continents was, very probably, greatly reinforced by variations in the strength of the sun's radiation. Certainly, the earth, some 3 million years ago, became a very much colder planet. An Ice Age began which brought glaciers down as far south in Europe as the Midlands of England, which has varied in its bitterness several times, and which has not left us yet.

The existence of a ring of continents around the Arctic has had a great effect on its animal population and made it a very different place from Antarctica, for the lands served as corridors along which animals from warmer parts of the world could advance up towards the ice. So whereas no large land animals except man live near the South Pole, the North Pole is the hunting ground of one of the biggest of all carnivores, the polar bear.

This huge white animal is related to the grizzly and the black bears that live south of the Arctic Circle in Eurasia and North America. It is most effectively protected against the cold. Like so many other cold-climate creatures, it is considerably larger than its relations living in warmer lands. The hair of its coat is very long, particularly oily and virtually impenetrable to water at shallow depths. Hair also covers most of the soles of its feet which not only keeps the skin off the chilling ice but also gives the animal a good grip. During the summer, farther south, a polar bear may eat berries and catch lemmings, killing them with swift dabs of its huge forepaws. But its main prey are seals. It will stalk one, moving with near invisibility, its white body pressed low against the snow. It may spot one basking on a floe and while still some distance away, dive and bob up at the edge, blocking the seal's escape route to the sea. Sometimes it will wait beside a seal's breathing hole in the ice and, when one appears, smash its head against the edge of the ice with a sideways blow of its paw.

Seals swarm in the Arctic as they do in the Antarctic. One species, the harp seal, congregates at breeding time in hundreds of thousands on the ice floes. But penguins, so conspicuous an element in the Antarctic scene, are missing. There are other birds here, however, that are very like them – members of the auk family, guillemots, razorbills, puffins and auks themselves. They resemble penguins in many different ways. They form huge colonies at breeding time; they are black and white, for the most part;

Polar bear

they have an upright stance on land; and most of all, they are excellent underwater swimmers, moving in much the same way as penguins by flapping their wings and steering with their feet.

Their change from fliers to swimmers, however, is not yet as complete as the penguin's. They have not entirely lost their powers of flight, though their wings are not now very efficient and when they launch themselves in the air they do so with a great deal of frantic whirring. At one short period of the year they all become flightless, for instead of moulting their wing feathers a few at a time, like most birds, they lose them all at once. Then they go out to sea and sit about on the water in great flocks among the waves, more penguin-like than ever.

One member of the family, the great auk, did become totally flightless. The largest of them, it stood upright and was 75 centimetres tall. It was also black and white and so was very similar indeed to a penguin. In fact, it was the first owner of that name. The origin of the word is disputed, some saying that it comes from two Welsh words meaning 'white head'. The bird did certainly have two white patches on its head, but it never lived in Wales. More likely, the name was derived from a Latin word meaning fat, for the great auk had an excellent insulating layer of fat beneath its skin and was much hunted for it. So when travellers to the southern hemisphere saw very similar flightless birds, they called them penguins too. That name stuck to the southern birds, but not to the northern ones. In the end, the great auk lost not only its name but its existence. Being flightless, it could not easily escape men. The last was killed in 1844 on a small island off Iceland.

The rest of the auk family survived, perhaps, precisely because they never lost their powers of flight. They congregate on inaccessible cliff faces and on the top of isolated rock stacks, but none stand about in vast crowds on beaches or ice floes as penguins do, no doubt because of the presence of those mammalian hunters who came up from the south.

They include not only polar bears and Arctic foxes, but men. The Eskimos made their way up from the lands of northern Asia in early times. They are now better adapted physically to living in conditions of extreme cold than any other group of human beings. Although they are not very large in stature, their bodies have the proportions best suited to heat-conservation, being squat with a small amount of surface area for their volume. Their nostrils are narrower than those of many other races, which may help them to reduce the amount of warmth and moisture lost through the breath. They even have heavy protective pads of fat in just those places that a fully dressed Eskimo leaves exposed to the cold – on their cheeks and their eyelids.

Even Eskimos would not be able to survive in the Arctic were it not for the warm furs of animals. They use sealskin for mittens and boots, polar bear skin for trousers, caribou pelts and bird skins for tunics. They sew the seams so finely that they will not let in the water, and wear two pairs of both tunic and trousers, the inner with the fur inside next to their skin, the outer with the fur on the outside.

Traditionally, Eskimos made long journeys on the ice, living entirely by hunting

Guillemots, Orkney

seals. When they camped, they used the snow itself to provide them with shelter, cutting it into blocks with a long blade of bone and stacking them in a rising spiral which met at the top, to build an igloo. They sometimes even provided it with a window by replacing one snow block with another of translucent ice. Inside, a bench cut from the solid snow was covered with skins. Light came from oil lamps. The heat from the burning oil and their own bodies could raise the temperature as much as 15°C, sufficiently high for the occupants to remove their heavy skin clothes and relax, semi-naked, among their fur blankets.

Such a life was one of almost inconceivable privation. Now, western man has settled in the Arctic and has brought with him new materials and fuels, electric generators and nylon fabrics, petrol-fuelled sledges and long-range rifles with telescopic sights. So the dog-hauled sledge and the hand-thrown harpoon, the igloo and the hand-sewn fur clothes have been abandoned. Today, Eskimos do not go on such long hunting trips across the ice floes of the Arctic.

The glaciers flowing off Antarctica meet the sea and float over it forming a massive shelf. Periodically this breaks up, forming immense tabular icebergs, some as much as 100 kilometres across, which may drift for decades through the Antarctic seas before they are finally carried up to warmer waters and slowly melt. In the Arctic, the edge of the ice-cap lies in many places on land. There, in Greenland, Ellesmere Island and Spitzbergen, it forms snouts and cliffs of ice from which issue streams of melt water. South of its margin for hundreds of kilometres stretches a desolate waste of gravel and boulders, the crushed and shattered debris of rocks pushed ahead of the advancing ice during colder eras and now abandoned as it retreats. This is the tundra.

During the summer, the feeble sun may melt the surface of the ground, but only a metre or so below it remains frozen solid as it has since the beginning of the last Ice Age. The soil above this layer of permafrost thaws and freezes with the seasons. The contraction and expansion within the gravels create strange shapes. If frost strikes a patch of ground, turning the moisture within it into ice, the gravels will rise up into a slight dome and expand laterally. Big particles are moved faster than smaller ones by the frost so that the finer gravel remains in the centre while larger stones are shifted towards the periphery. Where several such frost-prone patches form in the same neighbourhood, their margins may meet. So polygonal shapes appear in the ground, sometimes a few centimetres wide, sometimes over 100 metres across, outlined with rims of quite large stones. Since the finer gravels in the middle are more suitable for plants, a green centre appears in these polygons and a whole stretch of the tundra seems to have been divided into strange garden plots. On slopes, this process creates not polygons but long stripes running for great distances down a hillside.

Elsewhere, the regular freezing and thawing may concentrate the underground water so that it heaves up to produce a pyramid, 100 metres high, called a pingo. This has the shape of a small volcano but it contains, instead of lava, cold blue ice.

As might be expected, lichens and mosses grow all over the tundra, but over a thousand different species of flowering plants manage to live there as well. None achieves

Eskimo polar-bear hunters

the dimensions of even a small bush. The searing winds prevent that. Some, nonetheless, are trees. The Arctic willow grows not vertically but horizontally along the ground. Big ones may be 5 metres long yet only a few centimetres high. Like all these cold-climate plants, they grow extremely slowly. One with a trunk a couple of centimetres in diameter may well be 400 or 500 years old, as its annual rings will testify. There are also stretches of low heathers, sedges and cotton grass. Many of the tundra plants are also found at high altitudes on the mountains of North America and Eurasia. Indeed, they may well have originated there, for these mountains were in existence long before the last Ice Age overtook the world and the tundra was formed.

During the long dark months of winter, snow covers much of the land and there are very few animals to be seen. Beneath the snow where it is much warmer than above, lemmings, small rodents half the size of guinea pigs, fat and dumpy with thick brown fur, tiny ears and the smallest of tails, are trotting along runways close to the surface of the ground, cropping the vegetation. Occasionally, a white arctic fox will hunt for them, digging deep in the snow and pouncing with stiff legs to try and force the animals out of their tunnels. Ermine, little white carnivores, are small enough to chase the lemmings along their own tunnels. A few white birds, ptarmigan, may be searching sheltered valleys where some berries or willow leaves may be found. Arctic hares burrow into the snow desperately trying to find some un-nibbled leaves. But existence is difficult and only the fittest will be able to survive.

Spring comes suddenly. Daily, the sun rises higher above the horizon. The sky becomes a little lighter and the air a little warmer. The snows begin to melt. The permafrost prevents the melt waters from seeping away and they lie on the surface forming bogs and lakes. The animals and plants are quick to respond to these newer gentler conditions. This respite from the frost will only last about eight weeks. There is no time to be lost.

Swiftly, the plants put out their flowers. The green alder is in such a hurry that instead of opening its catkins and leaves one after the other as members of its family do elsewhere it unfurls them both at the same time. The lemmings, now that their protective covering of snow has melted, appear out in the open. In the pools and lakes, insect eggs that have lain dormant throughout the winter begin to hatch and hordes of blackfly and mosquitoes are soon emerging. The air is filled with a threatening drone as millions of insects search eagerly for the warm mammalian blood they need before they can lay eggs in their turn.

Insects and lemmings, green shoots and water plants are all excellent food for one creature or another and hungry migrants come up from the south to feast on this brief bonanza. Squadrons of duck appear – pintail, scaup, teal and goldeneye – and feed greedily on the plants that burgeon in the shallow lakes. Gyrfalcons, ravens and snowy owls come up for the lemmings. Phalaropes and dunlins and turnstones fly in to collect insects and larvae. Foxes come with them, reckoning on meals of eggs and chicks to come. And huge herds of caribou plod up to graze on leaves and lichens.

Now the white animals that spent the winter here have moulted and changed colour.

Tundra polygons, Alaska

Fox and ptarmigan, ermine and arctic hare, hunters and hunted, all need camouflage equally and all have become brown creatures safely inconspicuous on the snow-free tundra. The ermine has changed into an animal more familiar to southern eyes, a stoat.

The visiting birds are beginning to breed, rearing their young on the rich supply of insects. It all has to be done at great speed to make sure that the youngsters will be big and strong enough to tackle the journey back when winter returns, but there is now almost continuous light throughout the twenty-four hours of each day, so the parent birds can collect food and feed their young round the clock.

Then, with equal suddenness, the summer is over. The sun sinks lower daily. The light dims and hard frost seals the land once more. The showers of rain turn to stinging sleet. The phalaropes are the first to leave, but soon all the bird visitors together with their young are starting off. The caribou assemble in long columns and, with heads down, trudge back across the whitening land. They, like so many of the summer visitors to the tundra, will find shelter from the winter blizzards in the great forests of pine, fir and hemlock that lie farther south.

Caribou on the summer tundra

THREE

THE NORTHERN FORESTS

The caribou herds, moving southwards across the Alaskan tundra in September, are fat and in good condition after a summer of feeding. Their young are with them, gamely keeping up with their parents. But they have a long journey ahead of them and the weather is worsening. Snow is already falling on the bleak treeless land. The sun, during the day, is still strong enough to melt the snow, but this hampers rather than helps the caribou, for the melt water freezes again at night and the surface of the ground becomes glazed with ice in places so thick that the caribou cannot break through it to reach the leaves and the lichen beneath. As the need to reach shelter grows more urgent, they may travel as much as 60 kilometres in a single day.

At last, after a week or so of persistent walking, the herd reaches the first trees. They are stunted and gnarled and stand singly or in small groups in sheltered folds of the land. Still the caribou continue south. Slowly the trees increase in size and number. Eventually, after a march that may have taken them across 1000 kilometres, the herd moves through tall trees and into real forest.

Here, things are easier. Although it is still extremely cold, the dense trees shelter the animals from the heat-sapping, lethal wind. And there is food. The snow beneath the dark branches does not melt and refreeze, so it remains soft and powdery and the caribou, with a few strokes of their hooves and a nudge or two from their muzzle, can clear it away to find vegetation beneath.

The forest they have entered is the largest tract of trees in the world. It forms a band, in places 2000 kilometres broad, which extends right round the globe wherever there is land. From the Pacific shores of Alaska it stretches eastwards across the whole width of North America to the Atlantic coast. In the other direction, across the narrow hiatus of the Bering Strait, it continues right across Siberia and into Scandinavia. From one end to the other, it measures some 10,000 kilometres.

The improvement in conditions between here and the tundra further north, that allows trees to grow, is a slight increase in light. Nearer the Pole, the summer is so short that a tree has not enough growing-time in a year to build a tall trunk, nor to produce leaves tough enough to withstand severe frost before winter. Here, however, there are

Coniferous forest, Finland

usually at least thirty days in the year when the light is adequate and the temperature rises to 10°C or above – and that is just enough to enable a tree to develop.

In other respects, however, conditions are still extremely harsh. Temperatures can drop to 40°C below zero, even lower than the coldest temperature recorded on the tundra. Heavy blizzards cover the ground with snowdrifts metres deep that may persist for over half the year. The extreme cold not only threatens to freeze the liquid within the trees' tissues, but denies them one of their essential supplies – water. Although it lies all over the forest as snow and ice, in this solid form it is locked away beyond the reach of plants. So the trees of the northern forests have to endure a drought as extreme as that faced by many a plant in a sun-baked desert.

The kind of leaf that can withstand these privations is typified by the pine needle. It is long and thin, so snow does not easily settle on it and weigh it down. It contains very little sap, so there is little liquid in it to freeze. It is dark in colour and thus absorbs the maximum amount of heat from the feeble sunshine. All green plants inevitably lose some water as part of the process of growth. They need to absorb carbon dioxide from the air and to get rid of one of their waste products, oxygen, and this they do through microscopic pores called stomata. In the course of this exchange of gases, some water vapour inevitably escapes. The pine needle, however, loses far less than most leaves. It has relatively few stomata and they are placed at the bottom of tiny pits which occur in regular lines along the bottom of a groove which runs the length of the needle. This groove holds a layer of still air directly above the stomata and diffusion of water vapour from them is therefore very low. In addition, water loss through the cell walls elsewhere on the leaf's surface is reduced to almost nothing by a thick waxy coat. And when the cold becomes so severe that the ground is deeply frozen and all water supplies to the roots are cut off, so that any evaporation from the leaves could be disastrous, the stomata themselves can be closed.

In some circumstances, even these water-conserving devices are not enough. The larch grows in areas that are not only extremely cold but on ground that is very dry indeed. No loss of moisture during the winter can be risked and the larch sheds its needles every autumn and enters a state of total inactivity. Elsewhere, however, the needle-leaves operate efficiently and economically the year round and their owners keep them for as long as seven years, renewing them a few at a time during the growing seasons. Retaining leaves in this way brings considerable advantages. The leaves are on the tree at the very beginning of spring ready to photosynthesise just as soon as there is sufficient light; and the tree does not have to expend precious energy each year to build all its leaves afresh.

These evergreen needle-bearing trees belong to one ancient group that produce their seeds within cones and that appeared on earth some 300 million years ago, long before the rest of the flowering plants. They include pine and spruce, hemlock and cedar, fir and cypress. The nature of their needle-leaves, imposed by the harsh climate, determines to a considerable degree the character of the entire forest community that lives around them. Because the needles are so waxy and resinous, they do not decompose

Crossbill in pine tree

easily. Bacterial activity is, in any case, at a very low level because of the cold. So when the needles eventually fall, they remain undecayed on the forest floor for many years, forming a thick springy mat. Since the nutrients they contain are not released by decay, the soil beneath the mat remains poor and acid. The trees themselves are only able to reclaim the nutrients lost in their fallen needles with the aid of fungi. The roots of conifers are shallow and stretch in an extensive network around the bole, close to the surface of the ground. They are surrounded by webs and tangles of filamentous fungal hairs that extend upwards, investing the needles, breaking them down into chemical substances that the trees can reabsorb. In return for this service, the fungi are thought to extract from the tree roots the sugars and other carbohydrates that they need but, because they lack chlorophyll, cannot synthesise for themselves.

This relationship between the fungi and coniferous trees is not as intimate as the partnership with algae that produces lichens. Nor is it as specific. As many as 119 species of fungi have been found associated with just one species of pine, and six or seven kinds may live at the same time on the roots of one individual tree. Nor is it as obligatory; but without the assistance of the fungi, the conifers grow very much more slowly.

The nature of the needles also limits to a considerable degree the nature of the animals that can live in the coniferous forest. The immense annual crop of leaves produced by any forest might be thought to provide food for vast armies of vegetarian creatures. The waxy resinous needles, however, are regarded by most creatures, except insects, as virtually inedible. Caribou will not touch them. Neither will small rodents. One or two birds, such as the capercaillie and the pine grosbeak, do eat them, but even they greatly prefer the young spring shoots while they are still tender and juicy.

The most widely relished food produced by the conifers is not their leaves but their seeds. Several birds are able to extract them from the cones. The crossbill, a member of the finch family, has an extraordinary beak to help it do so. The two mandibles do not meet at the top but cross one another. This enables the bird to prise and lever the protein-rich seeds from the tough covering of a cone. It works with great industry and may collect as many as a thousand seeds in a day. The nutcracker is a much bigger bird, a crow that grows to a length of about 30 centimetres, and it has a bill sufficiently powerful and large to be able to crush the cones with a straightforward bite. This releases the seeds which the bird not only eats there and then but often stores in cracks in trees for later consumption.

A few small mammals – squirrels, voles and lemmings – also feed on the seeds, burrowing down through the snow in order to find them. The other, bigger vegetarians of the forest – caribou, roe deer and moose – draw heavily on their reserves of fat built up during the summer, but get a little sustenance by stripping bark from the trees, from the moss and lichens that grow on them or from the comparatively scarce bushes that do manage to grow in the more open parts of the forest, on the banks of rivers or the margins of lakes.

The carnivores that hunt these vegetarians have to patrol great areas of forest in order to get enough meat. The lynx, a large thickly-furred cat, may range over 200 square

Lynx

kilometres. In this cold country, where conserving energy is of such importance, the balance between profit and loss in hunting has to be assessed accurately. If a lynx gives chase to an arctic hare and fails to catch it within a zig-zagging sprint of 200 metres, it gives up. The energy expended will be likely to be more than that supplied by the meat of the hare's body. Roe deer, being so much bigger, are a much more worthwhile prey. Even a long hunt after them could be profitable and the lynx pursues the deer with great persistence. The wolverine, a carnivore the size of a large low-slung badger, will also tackle deer. For it the chase is often easier, for the thin crust of the snow allows it to run swiftly over the surface, whereas the deer, if they run into a drift, will break through, flounder and so get caught.

It is no surprise that many of these creatures living above the snow, where it is so much colder than beneath it, are giants of their kind. The capercaillie is the largest of grouse, the moose the largest of deer and the wolverine the largest of the weasel family. Their bulk aids them in conserving heat, just as it does the large mountain animals. But in number and variety, the animals of the silent frozen forests are few, and the snow may stretch for long distances unmarked by the tracks of any living thing.

This restricted community of animals and trees has much the same character across the entire spread of the coniferous forest. Indeed, if you landed by parachute in one part of it in the middle of winter, you would have to be a very good naturalist to determine from the animals you might see what continent you were on. The huge deer with immense antlers peering at you over its long drooping upper lip might be called in America a moose and in Europe an elk, but though the names are different, the species is the same. The smaller deer, sheltering in the forest for the winter, could be a caribou in North America or a reindeer in Europe, but again the two animals are virtually identical. Wolverines hunt in Scandinavia and Siberia as they do in North America. A small weasel-like creature with long glossy fur, raiding birds' nests, might be a European pine marten or an American marten, which is more compact and heavier, but only marginally so. And that most spectacular bird, the great grey owl, its feet warmly encased in feathers, flies all through the forests of both continents.

Other birds might be of a little more help to you for, although one species of crossbill inhabits the whole extent of the forest from east to west, there are several different species of nutcracker. The one in America has a grey body with black wings carrying white patches, whereas the European species is speckled. Sighting a black grouse-like bird nibbling the pine needles up in the branches could also be helpful. If it is as big as a turkey, then it will be a capercaillie and you will be somewhere in Eurasia from Scandinavia to Siberia. But if it is as small as a chicken and has a red eyebrow strip, then it will be a spruce grouse and you will be in North America.

When spring comes, however, the northern forests change their character dramatically. As the length of the days increases, the conifers make use of the additional light to put on a spurt of growth. Throughout the winter, the buds have been heavily muffled and sealed against the weather. A coat of resin has prevented loss of moisture; the cells on the outside have developed a kind of antifreeze in them that can endure temperatures

Capercaillie displaying

of minus 20°C without solidifying; and they are further wrapped by an outer layer of insulating dead tissue. Now the buds burst into life, splitting and pushing aside their winter coverings. Insect eggs, that have lain dormant within the needles, or stacked in holes bored beneath the bark by their industrious parents the previous summer, now hatch and ravenous armies of caterpillars emerge to feed on the young pine needles.

The caterpillars and grubs will themselves be prey to other creatures. They have two very different ways of protecting themselves. The caterpillars of the pine beauty moth, whose main enemies are birds, are dark green and so closely match the colour of pine needles that they are extremely difficult to spot. When they hatch, they distribute themselves widely through the branches so that the discovery of one does not lead immediately to the finding of many others on the same stem. The pine sawfly caterpillars, on the other hand, congregate in large numbers. Thousands may cluster on a single branch. Their main predators are ants which, given the chance, will seize the juicy caterpillars and carry them down the trunk to their nest. To find the sawfly caterpillars, the ants send out scouts. When one finds a swarm, it scurries back to its nest, leaving a trail of scent behind it. Platoons of worker ants can then follow the scent back to the caterpillars.

The sawfly caterpillars have no large jaws or poisonous stings with which to attack a scout ant, but they do have a method of preventing it from spreading the news of their presence. They collect the resin that exudes from the severed pine needles, chew it up and store it in a special pouch of their gut. When an ant scout discovers them, they dab this gum on its head and antennae. This so disorients it that it has great difficulty in finding its way back to base. Furthermore, the caterpillars have added to the gum a compound that seems to be very similar to the substance the ants themselves release as a danger signal. So if workers happen to come across the returning scout's trail, the scent does not stimulate them to follow it but warns them to keep away. Finally, if the hapless scout does manage to get all the way back to the nest, it gives off this danger signal so powerfully that the workers treat it as an enemy and kill it. So the great herds of sawfly caterpillars munch on undetected.

The trees now produce their flowers. Some are female – small unobtrusive tufts, often red and often borne at the tips of the shoots. The male flowers develop separately and produce such immense quantities of pollen that the forest is filled with drifting yellow mists. So the flowers are fertilised. But the summer will be so short that many species will not have sufficient time to develop their seeds. That will have to wait until next year. Those of the previous year, a little farther back on the stem, are only now beginning to do so, swelling to form green cones. Even farther back hang the brown three-year-old cones which have opened their woody scales and released their seeds.

Down on the ground, the lemmings and the voles that had spent the winter out of sight beneath the snow are scampering over the mat of needles, feasting on the fallen seeds. And breeding. A female lemming can produce as many as twelve young in a litter. She may do so three times in a season and young born in the first or even the

Great grey owl, Scandinavia

second litter can themselves breed before winter sets in, mating when they are only 19 days old and giving birth 20 days later. So soon the forest floor is alive with them.

How quickly the young mature and how many offspring they in turn produce depends on how much food is available. The abundance of different foods is not always the same. Every three or four years, for example, the trees will produce seeds in exceptional quantities. This may be due to variations in the warmth of summers, or to a need by the trees to accumulate food reserves over several seasons in order to produce a bumper crop. It may also be a positive adaptation by the trees to ensure the survival of their seeds. In normal years, lemmings and other seed-eaters consume such a high proportion of the seeds that few remain to germinate. In a year of exceptional abundance, so many seeds are produced that plenty are able to sprout before the lemming population has built up sufficiently to collect them. In the following year, the lemmings will not feed so well, produce smaller litters and their population will fall again.

The sprouting needles, the swarming caterpillars, the hordes of lemmings and voles, all are potential food, and as spring changes to summer, flocks of birds fly up from southern lands to gather the harvest. Owls arrive and join resident species to pounce on lemmings. Flocks of redwings, fieldfares and other thrushes come to feast on caterpillars; warblers and tits to pick off the adult insects. Now a human wanderer will have much less difficulty in deciding which continent he is on, for each section of the great evergreen forest, European, Asiatic or American, has distinctive birds from its warmer territories farther south. In Scandinavia there are bramblings and redwings; in North America, flocks of tiny yellow-flecked wood warblers of a dozen different species.

These visitors will spend the summer up here, taking advantage of the short-lived plenty to nest and rear their young. Their success in doing so will depend on the prodigality of the season, for the amount of food available each year varies greatly. Pine trees are not the only organisms to vary greatly in their productivity. The numbers of lemmings and voles also changes from year to year, gradually increasing over a period of five or six years and then falling catastrophically. This in turn affects the populations of owls that feed on them. In years when there are comparatively few voles, great grey owls, which feed almost entirely upon them, may lay only one or two eggs. But as the voles increase in the following year, so the owls, better fed, with more food resources to spare with which to manufacture eggs, produce bigger and bigger clutches. Eventually a year will come when the owls are laying seven, eight or even nine eggs in a clutch. But then the vole population crashes. The large population of owls faces starvation and, suddenly, a mass movement starts as they leave the northern forests and come south in a desperate search for food.

Similarly, crossbills, breeding rapidly in a year of abundant pine cones, will be compelled the following year when the pine crop is poor to travel south, where many of them will certainly die in the absence of the one kind of food they need.

The visiting warblers and tits and thrushes that have come north for the summer are only a part of the total population of their species on the continents. Other individuals have remained behind to raise their young in the gentler forests further south.

In these southern woodlands, the conifers are no longer dominant. As the climate becomes kinder, they are replaced first by birches then by an increasing variety of trees – oaks and beeches, chestnuts, ashes and elms. Their leaves are not tufts of dark needles but broad, wafer-thin constructions that they spread, tier upon tier, to catch the sunshine. The surface of these leaves is thickly covered with stomata, as many as 20,000 in one square centimetre. Through these, the trees absorb great quantities of carbon dioxide and manufacture the food they need to thicken their trunks and extend their branches. The amount of moisture that evaporates through their open stomata in this process is enormous. The water lost by a full-grown oak from the surface of its leaves in a single summer's day is to be measured in tonnes. But this presents no problem to the broad-leaved trees, for in much of these temperate lands, rain falls intermittently throughout the summer and there is no great water shortage in the ground.

These succulent broad bright-green leaves are much more palatable than pine needles and a great variety of animals feed on them. Caterpillars of all kinds swarm over them, each species favouring its own particular species of tree. Many feed at night when they cannot be seen by hungry birds. Others that are active during the day protect themselves with bristly poisonous hairs that birds find distasteful, and to prevent themselves being pointlessly killed, they advertise the fact with bright colours. Still others rely on camouflage and render themselves invisible by taking the colour of either the leaf they are destroying or the twig to which they are clinging. The match is so close that the best chance you have of finding them is to look, not for the insects themselves, but for the mutilated leaves they leave behind them. Hunting birds, it seems, adopt the same technique. At any rate, many caterpillars go to considerable trouble to dispose of their left-overs, carefully snipping off damaged stalks or partially-eaten leaves and letting them fall to the ground. Others take care not to rest after meals near where they have fed but clamber away to a distant twig.

The trees are not totally defenceless against this onslaught. They can develop chemicals in their leaves, such as tannins, that are so distasteful that many caterpillars will not eat them. Like any other defensive system, this can be expensive, consuming a significant part of a tree's gross product which could otherwise be used for constructive purposes such as building twigs and leaves. So these repellents are not made when they are not needed, nor when the insect attacks are on only a small scale. But when there is a major invasion, a tree such as an oak may swiftly manufacture tannins in those of its leaves that are under attack. This does not kill the caterpillars directly, but it induces them to clamber away to try and find better-tasting leaves elsewhere on the tree. In the process, they expose themselves to birds which are themselves in search of a meal, with the result that the number of caterpillars attacking the tree is significantly reduced. If the caterpillar infestation is really severe, one tree may actually warn its neighbours of impending trouble by releasing special messenger chemicals, undetectable to the human nose but registered by another tree which, in response, will begin to produce tannins in its own leaves, even before the caterpillars reach it.

One family of birds, the woodpeckers, is particularly well-adapted to living in forests.

Their feet are modified for clinging on to vertical tree trunks, with their first and fourth toe pointing backwards and the second and third forwards. The feathers of their tails are short, thick-quilled, and stiff, so that the tail as a whole serves as a prop. And their bills are sharply pointed, like a pick. They cling upright on the trunk, listening intently for the tiny sounds that will give away the presence of an insect moving within its tunnel in the bark. When they detect one, they hammer away with their beak, opening up the insect's gallery and flicking their tongue into it to collect their prey. This tongue has a barb at the end and is extraordinarily, almost unbelievably long. In some species it is as long as its owner's body and is housed in a sheath within the skull that curls round the eye-socket and ends in the base of the upper mandible of the beak.

The birds also use their powerful beaks to excavate nest holes in tree trunks, drilling first of all a neat horizontal hole, then chiselling downwards for a foot or so and there cutting out a chamber. They frequently choose dead trees, no doubt because the rotting wood is softer to work than that of living trees. Such trees, also, are usually infected by bark beetles which provide a rich food supply conveniently near at hand.

The drumming noise made by the rapid blows of a woodpecker's beak on a tree trunk is one of the most characteristic sounds of the forest. The birds do not produce it solely when they are feeding or excavating a nest. They beat tattoos on resonating timber for the same reason that other birds sing, to declare possession of a territory and to attract a mate. Each species has its own length of drum roll with its own characteristic interval between one burst and the next.

Different species of woodpecker specialise in different foods. The green woodpecker, as well as taking bark-boring beetles, often descends to the ground to forage for ants. The wryneck is even more dependent upon them. It is not primarily a climber at all and lacks the stiff propping tail of other woodpeckers, but it does have the long sticky tongue which it flicks into a nest of ants to bring out 150 of them at a time. The acorn woodpecker exploits its wood-boring skills by drilling neat sockets in tree trunks, the diameter of which exactly accommodates acorns. It will cover a favoured tree with several hundred such holes and store several acorns in each of them, so accumulating a massive larder for the winter. An even more specialised group within the family, the sap-suckers, bore holes in tree trunks for a quite different purpose. They choose living trees of species that produce liberal flows of sap and drill numerous small, squarish holes in them. The liquid that trickles out, resinous or sweet according to the kind of tree, attracts insects which the sap-sucker collects and then mixes with the sap to produce a little sweetmeat, rich in both sugars and proteins.

As the warm days pass, the broad-leaved trees produce their flowers. The woods are not so dense or so tall that they exclude all wind and most trees rely upon it to carry their pollen to the waiting female structures. So the flowers are for the most part small and unobtrusive, since they have no need to attract insect couriers to carry the pollen. Unlike the summer farther north, the summer here is long enough for the flowers to develop their seeds within the same season. So chestnuts swell and acorns appear; sycamores produce sheaves of their winged seeds and hazels their hard-shelled nuts.

Yellow-bellied sap-sucker

Summer nears its end. The days begin to shorten giving warning of the coming cold. The trees now prepare themselves for winter. If the leaves, so thin and so full of sap, were to remain on the branches, they would certainly be damaged by frost. The winds of winter might well catch them so strongly in a gale that a whole branch could be torn off. Nor could they function very effectively in the shorter dimmer winter days, yet they could still lose precious moisture through evaporation from the stomata. So the leaves are shed. First the green chlorophyll in them is broken down chemically and withdrawn. This unmasks the waste products of photosynthesis and the leaves turn brown, yellow and even red. The vessels that carried sap to and from the blade of the leaf develop blockages at the base of the stalk, sealing them off, and a band of corky cells appears in the same place. Soon the slightest wind is enough to detach the dry leaf from the branch. So begins the autumn and the fall.

Many of the mammals of the broad-leaved forest – shrews and voles, mice and squirrels, weasels and badgers – will have to survive the approaching winter on a much reduced diet. They do so by absorbing the fat reserves that they accumulated during the summer. They also reduce their activities to a minimum, avoiding any unnecessary expenditure of energy and spending a great deal of their time within their holes and burrows. Others follow the policy of the trees and enter a state of suspended animation. The depth of their winter slumber varies. Black bears are among the lighter sleepers. In the early autumn, they look for holes among rocks, leaf-filled hollows hidden under an overhang, or caves. Often the den they choose is one they have slept in many times before. Each animal beds down by itself. There, after a month or so of drowsiness, the female gives birth to her cubs, usually two or three of them. She seems scarcely to notice their arrival, for they are extraordinarily small, no bigger than rats. They snuggle into her fur as she slumbers and find their way to her nipples. The mother herself neither feeds nor urinates nor defecates, nor will she until spring.

As the winter passes, the cubs grow rapidly. They growl and whimper like puppies as they move blindly around in the dark den, on occasion making such a noise that even if you are many yards away, you may hear a strange murmuring hubbub in what appears to be a completely empty tract of frozen wilderness. The length of time the mother and her cubs remain in the den depends on how long and how severe the winter is. In the southern part of the American woods, their retirement may last little more than four months, but bears in the northern parts may spend six or even seven months in their dens and so pass the greater part of their lives in a doze.

During this sleep, the bear's heartbeat slows and its body temperature falls a few degrees. This saves the bear valuable energy yet allows it to wake up quickly if disturbed.

Smaller creatures, like dormice, hedgehogs and woodchucks, on the other hand, fall into a sleep so profound that it can be difficult to be certain that they are still alive. One will curl up, head tucked into its belly, hind feet near the nose, fists clenched and eyes tightly shut. Its body temperature falls to only a degree or so above freezing and its muscles set rigidly, so that the animal not only feels as cold as a stone but seems, beneath its fur, to be almost as hard as one. In this state, the bodily processes only

Black bear hibernating

barely tick over, making the minimum demands on the animal's reserves of fat. During the summer, a woodchuck's heart beats about 80 times a minute. In winter that rate drops to four, and instead of taking 28 breaths in that period, it takes only one.

But this death-like dormancy does not necessarily last throughout the winter. A warming of the weather may wake the animal up. More surprisingly the opposite stimulus does the same, for if frost were to penetrate its refuge and chill it by one more degree, it will freeze and die. So if there is a spell of intense cold the sleepers rouse themselves, and reactivate their internal batteries to save their lives, expensive though that will be as far as their fat store is concerned. Dormice and marmots prepare for this eventuality by laying up stores of nuts and other foods, either in their sleeping quarters or nearby, and have a quick snack. As soon as the extreme conditions slacken a little, they return to their shelters and go back to sleep.

The trees are now bare. Their leaves, carpeting the woodland floor, are decaying rapidly. Though it is cold, the ground is only frozen for short periods and bacteria and fungi can get to work. Other small creatures – beetles and millipedes, springtails and, particularly, earthworms – churn through the leaf litter, mixing it with the earth, turning it into a rich humus-laden soil. Within two years of falling, nearly all the broad leaves will have completely disintegrated. Pine needles, even in these mild conditions, are likely to take more than twice as long.

Farther south still, the winter withdrawal from active life is not so extreme. The cold is never so severe that the trees risk having their leaves crippled by frost and many species of broad-leaved tree appear – magnolias, olives and arbutus – that are evergreen. Species of families that farther north shed their leaves, such as oaks, here keep them throughout the year. The most taxing time for trees in these regions is not in winter but in summer when the weather becomes so hot that the trees are in danger of losing their liquids. The leaves of many of these broad-leaved evergreens, therefore, are usually dry, with a waxy watertight surface and relatively few stomata, often mainly on the underside. Many hang downwards from the branches during the hotter part of the day so that they do not catch too much of the sun's heat. When you sit under them you discover that, in consequence, they give surprisingly little shade.

And here, the conifers reappear. The very characters that enabled them to survive the water shortages caused by freezing temperatures serve them just as well in the hot southern summers. Their shape, however, is different. In the north, many were pyramidal with branches that sloped down and away from the trunk, helping snow to slide off them and so preventing the accumulation of heavy loads that might break them. Here that danger is not so great and the trees can stretch their needle-clad branches upwards and outwards to gather the maximum amount of light. So the typical conifer of the south is the flat-topped, expansively branching umbrella pine.

The conifers' techniques of water conservation are so effective that they can grow on soils that are too well drained, too sandy and dry to allow broad-leaved trees to survive. But in some parts of the south, pines maintain their position on ground that is sufficiently well-watered and fertile to be considered broad-leaf territory. They survive in such places because of another of their talents, their ability to endure fire.

In the southern United States, in Florida and Georgia, the long hot summers bring heavy and regular thunderstorms. The immense clouds, towering several miles high, shed torrential rain and discharge bolts of lightning which strike the taller trees, burning a zig-zag groove down the trunks as they go to earth and sometimes even splitting the tree. Often these lightning strikes set fire to the leaf litter on the ground and flames sweep swiftly through the forest. Pines have a pulpy bark that scorches but does not burn and keeps much of the heat away from the sensitive tissues beneath. The buds at the tips of the stems of the young plant, well within reach of the flames, are surrounded by a thick tuft of long needles which does burn but at a relatively low temperature that does not harm the bud. By the time they have been consumed, the main fire has swept by. The young oak seedling has neither of these protections. When the flames, fuelled by the leaf litter, lick round its infant trunk, they, in effect, cook the growing cells just beneath the thin bark and strike directly at its unprotected buds, killing the plant within a few minutes. So the young broad-leaved trees die and the young conifers survive.

Conifers not only tolerate these conditions; to a considerable degree they create them. The resinous needles they shed, so resistant to decay, form an excellent tinder and lightning is more likely to start fires in a coniferous forest than in a broad-leaved one. Conifers positively benefit from the fires. The flames not only exterminate competing plants, they release the nutrients in the needles and reinvigorate the soil. The smoke also kills fungi that might attack the trees. Some pines even produce cones covered with a pitch-like resin that will only open to release their seeds after being subjected to intense heat. Protecting such forests as these from fire, and maintaining teams to snuff out any fire that might start, is to interfere with a natural regime and is likely, in the long run, to change the forest's character, turning it from one dominated by conifers into a broad-leaved one. But it can also lead to great danger.

Without regular fires, fallen leaves, branches and dead trunks slowly accumulate on the ground. If, after many years of fire suppression, one finally starts and gets out of control, all this dry fallen debris catches light. Fires of this kind may stay burning in one area for hours. As they grow in size, a fire storm develops, the flames roar up the trunks, and the crowns of the trees explode into fire balls. No tree can survive such burning and the forest is destroyed.

In normal circumstances, the frequent fast-moving fires cause little trouble to the animals. Birds can fly away from them. Ground-living creatures such as rattlesnakes and gopher tortoises take refuge for the few minutes that the fire takes to sweep by in the holes they regularly use as shelter from the oppressive midday heat. Rats and rabbits have been seen watching the advancing line of flames, selecting a section where they are relatively low, and deliberately running through the thin line of heat to safety on the blackened ground on the other side.

The woodpeckers of the southern forests, however, are in real jeopardy. Were they, like their northern cousins, to excavate their nest holes in the trunks of dead trees, the fire could easily get a hold on the trunk and asphyxiate or even incinerate their young. The typical woodpecker of these forests, the red-cockaded woodpecker, avoids these dangers by chiselling its hole not in dead conifers, but living ones.

Doing this, however, causes new problems. Conifers protect themselves from injury to their trunks and branches with resin. A branch torn off by a gale, a tunnel bored by an insect, a cut delivered by a forester's axe, will quickly exude a strongly smelling fluid, which hardens in the air and forms a scab over the tree's wound, keeping its precious sap in and infection out. This resin is carried in vessels that run up the tree in the outer layers of the trunk. Were the woodpecker to cut its nest hole in this part of the living trunk, the quantity of resin flowing into the nest would make the hole uninhabitable because of its resinous vapours and its sheer stickiness. So the red-cockaded woodpecker bores right through this zone of the trunk and into the heartwood. For the trunk to contain sufficient heartwood to accommodate a woodpecker's chamber, it has to be very thick. So usually the red-cockaded woodpecker nests comparatively low down on the trunk. This, however, puts it at risk of being plundered by other animals, in particular by rat snakes which regularly climb trunks and steal young chicks from their nests. The woodpecker has a defence against such raiders. It pecks a line of pits both above and below the entrance to the nest, very like those made by sap-suckers. These it tends regularly so that they exude resin in great quantities and form a sticky sheet right round the trunk. The chemicals in the resin seem to irritate the snake's belly unendurably and when a climbing snake encounters a patch, it suddenly rears back, writhing and arching its body, loses its grip and falls to the ground.

Holes, once made, may be used by the same bird year after year. Some serve as nests, others are used as roosts. But cutting a hole in a living tree is much more difficult than excavating one in the softening wood of a dead trunk. Holes are, therefore, very valuable property and many other creatures, mammals such as squirrels and birds such as owls, which lack the woodpecker's talent for carpentry, will commandeer them if they get a chance. The woodpeckers have to maintain a constant guard. The red-cockaded woodpecker has developed a solution to this problem too. The bird lives in clans of eight to ten individuals, all of whom are closely related. But only one pair will nest. The others, usually younger birds, take turns in guarding the occupied nest hole. They may even assist in feeding the young and certainly take turns in working on new nest holes.

These American pine woods form the southern edge of the great band of forest that once, before man felled so much of it, stretched right across the continents of the northern hemisphere. Throughout all this vast area, the plants and the animals have to cope with conditions that vary widely during the course of a single year and which, at some times, can be very harsh indeed. Those living in the northern parts have to be able to deal with days that are light for nearly all their length as well as others that are almost totally dark; those in the south with days of cold drenching rain, and then, later in the year, weeks of desiccating sunshine. To cope with these varying conditions, animals and plants have had to develop particular strategies and structures. None of them is able to operate continuously at its highest potential efficiency the whole year round.

But less than 1000 kilometres further south towards the equator lies the Tropic of Cancer. Beyond that imaginary line, the sun stands, in due season, directly overhead. Here there are lands where it shines brightly almost every day, where frost never strikes and where life-giving rain falls almost every day. Such a country was the original home of the broad-leaved trees and there they still dominate the land and reach their greatest development. Indeed, life of all kinds flourishes there in greater abundance than anywhere else in the world.

FOUR

JUNGLE

Nowhere is there more light, warmth and moisture than in West Africa, Southeast Asia and the islands of the western Pacific, and South America from Panama across the Amazon basin to southern Brazil. As a consequence, these lands are blanketed by the densest and richest proliferation of plants to be found anywhere in the world. Technically it is described as evergreen rain forest. It is more widely known as the jungle.

Conditions within it, compared to those within the forests farther north, change hardly at all. Being close to the equator, the amount of sunshine and the length of the day remains almost the same throughout the year. The only variation in the rainfall is a marginal one – from wet to somewhat wetter. It has been like this for so long that, in comparison, all other environments except the ocean, seem merely temporary phases. Lakes fill with mud and turn to swamps inside decades; plains turn to deserts inside centuries; even mountains are worn down by glaciers within millennia. But hot, humid, jungle has been standing on the lands around the world's equator for tens of millions of years.

This very stability may be one of the causes of the almost unbelievable diversity of life that exists there today. The great forest trees are far more varied than their uniform smooth trunks and nearly identical spear-shaped leaves suggests. Only when they produce their flowers is it evident how many different species there are among them. The numbers are astounding. In one hectare of jungle, it is common to find well over a hundred different kinds of tall tree. And this richness is not restricted to plants. Over 1,600 species of bird live in the jungles of the Amazon and the number of insects there is almost beyond computation. Entomologists in Panama collected from just one species of tree over 950 different species of beetle alone. Considering insects as a whole, together with other small invertebrates such as spiders and millipedes, scientists estimate that there may be as many as 40,000 different species in a single hectare of South American forest. It seems as though the processes of evolution, working uninterruptedly in this stable environment for so many millions of years, have produced specialised organisms to suit every tiny niche.

The majority of these creatures, however, live in a part of the jungle that, until quite

Jungle canopy from the air, Peru

recently, was largely beyond our reach and virtually unexplored at close quarters – the canopy of leaves, 40 or 50 metres above the ground. That an abundance of creatures live up there is evident to anyone, if only because of the extraordinary variety of clicks and whirrs, howls and screams, trills and coughs that echo down from the branches by day and, particularly, by night. But exactly what creature makes which noise is often largely a matter of guesswork. An ornithologist, with binoculars to his eyes, craning his neck upwards, is lucky if he sees more than a silhouette flitting briefly across a gap in the branches. Botanists, baffled by the uniformity of the huge pillar-like trunks and unable to examine the growing flowers, have been driven to shooting down branches with a shot-gun in order to identify the trees around them. One worker, determined to compile as complete a catalogue as possible of the trees in the jungle of Borneo, even trained a monkey to clamber up selected trees, pluck sprigs of flowers and throw them down to the ground.

Then, a few years ago, someone brought to the jungle rope-climbing techniques that had originally been developed by mountaineers. The systematic first-hand exploration of the rain forest canopy had at last begun.

The method is simple. First, you must get a thin line over a high branch, either by throwing it or by attaching it to an arrow and shooting it. To this you tie a climbing rope, as thick as a finger and strong enough to hold many times a man's weight. Hauling on the line pulls the rope over the branch. Once it has been securely tied, you clip on to it two metal hand-grips. You can slide them upwards, but a ratchet prevents them from slipping down. With your feet in webbing stirrups, one tied to each grip, you can slowly inch your way up the rope, standing with your weight in one stirrup while you hoist the other. Slowly and exhaustingly you climb to one high branch, fix another rope to a branch high above and climb up that, until eventually you have a single long rope attached to one of the topmost branches. Now, at last, you can climb into the canopy.

Arriving there is like leaving the dim airless staircase of a tower block and emerging on to its roof. Suddenly the humid twilight is replaced by fresh air and sunshine. Around you stretches a limitless meadow of leaves, hillocked and dimpled like the surface of an enormously enlarged cauliflower. Here and there, standing 10 metres or more above all the rest, rises a single isolated giant tree. These emergents live in a different climate from the other jungle trees, for up here the wind blows freely through their crowns. They take advantage of the fact by using the wind to spread both their pollen and their seeds. The giant kapok, or silk cotton, of South America produces vast quantities of fluffy seeds that float like thistledown and pervade the forest for miles around. Its equivalents in Southeast Asia and Africa equip their seeds with wings so that their spinning fall is sufficiently slow to allow the wind to catch them and transport them long distances before they disappear into the canopy.

The wind, however, also brings disadvantages. It can rob a tree of one of its vital supplies by evaporating moisture from its leaves. The emergents have responded to this danger by producing narrow leaves that have a much smaller surface area than the

Giant forest tree, Peru

leaves of canopy trees or even those of their own leaves that sprout from the lower branches in the shade.

The crowns of these huge trees are the favoured nesting place for the most ferocious bird in the jungle, a giant eagle. Each forest has its own species – in Southeast Asia the monkey-eating eagle, in South America the harpy, in Africa the crowned eagle. They are remarkably similar. All have large crests, and broad relatively short wings with long tails, a shape which gives them great manoeuvrability in flight. They build a gigantic platform of twigs which they occupy season after season. On it they usually raise just a single nestling which remains dependent upon its parents for food for almost a year. And they all hunt with ferocity and at speed in the canopy. The harpy is, marginally, the largest of all eagles anywhere and will pursue monkeys, twisting and diving through the branches as the monkey troop flees in panic, until it finally pounces and carries its struggling victim back to its nest. There the corpse will remain for several days as the eagle family dismember and eat it, piece by piece.

The canopy itself, the ceiling of the jungle, is a dense continuous layer of greenery some 6 or 7 metres deep. Each leaf in it is accurately angled to ensure that it will collect the maximum amount of light. Many have a special joint at the base of the stalk that enables them to twist and follow the sun as it swings overhead from east to west each day. All except the topmost layer are largely screened from the wind, so the air around them is warm and humid. Conditions are so favourable to plant life that moss and algae grow abundantly. They coat the bark of the branches and hang from the twigs. Were they to grow on a leaf, they would deprive it of the sunshine it needs and clog the stomata through which it breathes. The leaves, however, are protected against this danger with glossy waxy surfaces on which a rootlet or a filament would have great difficulty in getting a hold. Furthermore, nearly all the leaves have drip tips, elegant spikes at their ends like tiny spouts, which ensure that after a rain storm, water does not linger but swills over the surface and drains swiftly away, so keeping the top of the leaf well washed and dried.

There are no well-defined seasons in the jungle and there is therefore no obvious climatic cue for all the trees to shed their leaves simultaneously, as there is in other latitudes. But this does not mean that all trees shed and regrow their leaves continuously throughout the year. Each species has its own timing. Some drop leaves every six months. Others do so after what seems to be a quite arbitrary period with no discernible logic in it – every twelve months and twenty-one days, for example. Still others do so piecemeal, at intervals throughout the year, a branch at a time.

Flowering periods also vary and do so even more extremely and mysteriously. Ten-month and fourteen-month cycles are common. Some trees, exceptionally, flower only once in a decade. Again, the process is not haphazard, for all the trees of one species over a vast area of jungle will burst into flower simultaneously, as they must do if they are to cross-fertilise one another; but the stimulus that triggers them all is still undiscovered.

The flowers produced by the canopy trees, unlike those of the emergent giants,

Monkey-eating eagle, Philippines

cannot rely on the wind for their pollination since the air around them is almost still. They must, therefore, attract animal messengers and this they do with nectar, advertising the fact that it is available for visitors with conspicuously coloured petals. Many are fertilised by insects, lumbering beetles, wasps and strong-winged brilliantly coloured butterflies. Those that rely on nectar-feeding birds – hummingbirds in South America, sunbirds in Asia and Africa – are nearly always red, whereas those that are pale and have a foetid smell are usually patronised by bats.

Similar transport problems occur when the seeds develop. Seeds are bigger than pollen grains so the creatures recruited for this job must be of a fair size. Many trees, therefore, wrap their seeds in succulent sweet flesh that attracts monkeys and hornbills, toucans and fruit bats – all creatures quite big enough to swallow seeds with their fruit without even noticing. Figs are eaten up on the branches. Bigger fruit – avocado, durian and jackfruit – drop to the floor where they are taken by ground-living creatures. In all these cases, the seed itself has a hard, tough coat so that it can pass along the entire length of an animal's digestive tract and emerge at the other end with the droppings, unharmed and, with luck, some distance away from where it was consumed.

A rich and varied community of animals lives in the green world of the canopy, high above the ground, browsing and hunting, thieving and scavenging, breeding and dying, without ever leaving it. With so many different species of tree fruiting at different times, there is usually fruit to be found somewhere or other throughout the year so it is possible for animals to become specialist fruit-feeders and eat very little else. They form a wandering gang of birds and mammals that move from one tree to another to plunder the fruit as soon as it becomes available. One of the most rewarding ways of watching canopy life is to find a tree that is just coming into fruit, and sit and wait. A fig tree in Borneo, if its fruit is ripe and fragrant, will be swarming with creatures. Monkeys scamper about in its branches sniffing every fig individually to decide from its perfume whether it has yet reached perfection, and then, if it is to their liking, cramming it into their mouths. That great red-haired ape, the orang utan, is by nature a solitary and usually there will be only one male or a female with her baby in the tree. Whole families of gibbons, however, turn up; and out on the furthermost, thinnest twigs, where heavy creatures find it difficult to move, fruit-eating birds flutter and squawk. Parrots clamber about, clasping the fruit in the claws of one foot while hanging upside down from the other; hornbills and toucans pick them off one at a time with their long beaks and throw them in the air to catch them at the back of their throat. Nor does the banquet stop at the end of the day. New customers arrive at night. Perhaps a loris, a nocturnal primitive primate, pale-furred and wide-eyed, will emerge from its hiding place, and giant fruit bats land on the branches with a rustle and flap of leathery wings.

Other creatures specialise in feeding on the vast and inexhaustible supply of leaves. Cellulose, however, is not easy to digest and animals that depend on it must have large stomachs in which to hold their meals while it is being broken down. In consequence, most leaf-eaters are quite big creatures and include very few birds, for whom weight has to be kept to a minimum if they are to remain capable of flying. A few monkeys have

Spider monkeys drinking from a canopy flower, Central America

adopted leaves as a major part of their diet and developed particularly large and specialised compartments in their guts in order to deal with them – howlers in South America, leaf monkeys in Asia and colobus in Africa. The oddest of all canopy leaf-feeders is surely a South American, the sloth. It lives suspended beneath branches, moving in a stately way along the boughs, shifting one foot at a time. Its claws have become hooks and its limbs transformed from supple jointed props to stiff bony hangers. Its hair tracts run the opposite way from normal creatures, from its ankle to its shoulder and from the middle of its abdomen towards its spine, so that rain falling on it as it hangs upside down easily drains away. The three-toed sloth tends to live in lower trees and feeds almost exclusively on leaves of the cecropia, but the two-toed sloth is a true canopy dweller, clambering about in the topmost branches and eating not only a wide variety of leaves but fruit as well.

Hunters are up there too. In addition to the great eagles plunging through the canopy to grab monkeys or birds, there are also tree-living cats – in South America the margay, in Asia the clouded leopard. Both are superbly athletic tree climbers, quite capable of stalking and catching monkeys, squirrels and birds. They leap from branch to branch, hang from their hind legs, and race up trunks. Their reflexes are so quick they can even, if they fall, catch hold of a passing branch and save themselves with one paw. There are also snakes here. Not the great monsters, so common in romantic fiction, that dangle optimistically from a branch, waiting to pick up a human passer-by, but much smaller creatures, some twig-thin, that collect frogs and nestling birds.

Many of the canopy dwellers claim their own territory, small or large, among the branches, which an individual, a family or even a troop will defend against intrusion by others of its own kind. Visual displays cannot be widely seen in the dense leaves and scent marks, so commonly used on the ground, are laborious to apply and maintain, and are not very effective among the tangled branches. Sound signals are much easier to send and travel much farther, and canopy animals produce some of the loudest of all animal noises. Howler monkeys, morning and evening, indulge in choral singing producing eerie wails, rising and falling for minutes on end. Male and female gibbons sing long duets of cascading calls, their contributions fitting together so perfectly that it is easy to assume that there is only one singer. In the Amazon forest, the bell bird, a pure white creature hardly bigger than a thrush, sits all day in the top of the trees reproducing the sound of a cracked anvil being clouted with a hammer so piercingly and so insistently that it can drive human travellers to distraction.

The scaffolding of massive boughs grown by the trees to support their own leaves is also used by other plants. Tiny spores of ferns and mosses, floating through the air, often become lodged in the crevices of the bark, and there they sprout. As they flourish and decay, their remains form a compost that is capable of supporting bigger plants, so that as a tree ages its broad boughs acquire a line of huge ferns, orchids and bromeliads, drawing their sustenance from the leaf mould accumulated on the branch and collecting their moisture by dangling their roots into the humid air.

The bromeliads, in turn, support their own tiny community of lodgers. Their leaves

Two-toed sloth

grow in a rosette, the bases clasping one another so tightly that they form a chalice that holds water. To these miniature ponds come brilliantly coloured frogs. Their eggs are not laid here, but usually on a leaf. When the tadpoles hatch, the female allows them, one at a time, to crawl up on to her back. Then she makes her way to a bromeliad. On arrival, she inspects it carefully, staring fixedly at the water-filled axil of one of the leaves. If there is no sign of life in it, she carefully backs down until her rear touches the water and her tadpole can wriggle into its own special aquarium. Several species of small frog behave in this way, relying on mosquitoes and other insects to lay their eggs in the bromeliad as well and so provide the developing tadpole with its food. But one species cares for its young in an even more elaborate way. The female visits each of her several young once or twice a week and deposits in the water alongside each of them a single infertile egg. The tadpole quickly bites through the jelly and begins to feed on the yolk. For six to eight weeks the female provides for her young in this way, until at last they develop their legs and take up independent life.

Not all the plants on the boughs of the trees are mere squatters, like the bromeliad. Others are more sinister. Fig seeds often germinate here, but their roots do not remain dangling innocuously in the air, like those of a bromeliad. They continue to grow downwards until they reach the ground. Then they dig into the soil and begin to absorb more water and more sustenance than they were able to find in the air. Their leaves, up on the bough, in consequence begin to grow more vigorously. Other roots grow along the bough, or horizontally out from the dangling roots, and begin to wrap themselves around the main trunk of their host. The crown now burgeons so vigorously and produces such dense foliage that it begins to overshadow the host tree. Slowly the fig becomes more and more dominant. Eventually, maybe a century after the fig seedling first sprouted on its bough, the host tree, having lost its share of the light, dies. Its trunk decays, but the fig roots that enwrapped it are now so thick and substantial that they form a hollow trellised cylinder, quite able to stand by itself. So the strangler fig has usurped its host and taken over command of this position in the canopy.

Other less dangerous stems, the lianas, climb up the canopy trees. They start life on the ground as small shrubs but they put out numerous tendrils with which they grope for sapling trees. If they find one, they cling to it and as the sapling grows so the liana is carried upwards until the two reach the canopy together. But the liana keeps its roots in the ground and takes nothing from the tree except support.

So lianas, strangler figs and the hanging roots of bromeliads and ferns festoon the canopy trees, like ropes around a ship's mast. If you have climbed into the canopy yourself, yours will be dangling among them. Descending it is not difficult, though it does require a certain confidence in your ability to tie knots correctly. A loop of rope is put through and around a metal figure-of-eight ring. You clip this to your waist belt. Standing in your webbing stirrups you can now slide downwards, controlling the run of the rope, and therefore your speed, with your hand. After you have dropped 10 metres or so, you have cleared the canopy branches and will be swinging free with nothing around you but the lianas and roots and, beyond, the monumental unbranched trunks

Arrow-poison frog carrying her tadpole

of the canopy trees, rising smooth and massive like columns in a Norman cathedral. It might seem that in this emptiness, between the green ceiling above and the ground beneath, there would be little to see. But there is a lot of traffic through this air-space as creatures make their way between ground and canopy. Some, as you are doing, use ropes. Squirrels run up the lianas. Orang utan, which when adult sometimes get so big that they find it difficult to cross from one tree to another in the branches, often descend to the ground and climb up another liana, ascending hand-over-hand with enviable ease. Sloths which, somewhat surprisingly, always defecate on the ground, and usually in the same place, may be clambering slowly downwards on their way to visit their midden.

Many birds prefer to travel from one part of the forest to another beneath the canopy, rather than expose themselves to the eagles patrolling it above. Many nest here. Macaws, hornbills and toucans use holes in hollow trees; trogons excavate a chamber in the globular nests of tree ants; tree swifts stick together bits of bark and feathers with spittle and build a tiny extension on the side of a horizontal branch in which their single egg fits as snugly, and with as little to spare around the edges, as an acorn in its cup.

Birds are not the only creatures that may sail past you through the air. Other kinds of animal, while they cannot flap their wings and therefore do not have powered flight, are nonetheless very competent aeronauts. They glide. Borneo is particularly rich in them. Among the squirrels running up the trunks and along the branches, clinging securely to the bark with their needle-sharp claws, is a particularly large and handsome one, a rich russet-red both above and below. You are most likely to see it in the late afternoon, when it emerges from a hole in a tree. Usually, a second will follow, for they live in pairs. For a minute or two they will circle the trunk and then, suddenly and unexpectedly, one will leap off, unfurling as it does so a great cloak of skin that stretches on either side from wrist to ankle. Its long bushy tail streams out behind acting, it seems, like a rudder. As one goes, its partner is likely to follow and the pair sail past you for maybe 30 or 40 metres towards another trunk. As they near it, they swoop upwards, so slowing their glide and enabling them to land, head uppermost, on the trunk, up which they then gallop, their furry membranes flopping around them like outsized overcoats.

A small lizard also rockets from liana to liana and branch to branch. Its gliding membrane is not all-enveloping like the flying squirrel's but merely a flap of skin projecting from either flank, stiffened by bony elongations of its ribs. Normally it keeps these flaps furled alongside its body, but when it pulls its ribs forward, they spread apart and so distend the flaps. These little creatures are extremely jealous of their own territories among the branches. If another intrudes, the owner will immediately dive into the air and land quite close to its rival. There it begins a furious aggressive display, flicking out a triangular flap of skin beneath its chin until eventually the intruder is outfaced, scuttles down the branch and glides away.

A few frogs have also taken to gliding. They use the membranes between their toes that are part of a standard frog's swimming equipment. The flying frog has greatly elongated toes and when they are extended, each foot becomes in effect a tiny parachute,

Top: *Flying squirrel*
Bottom: *Flying lizard*

so that when the frog leaps it can glide considerable distances from one tree to another.

Perhaps the most extraordinary glider of all and one whose exploits were, for some time, dismissed as the fantasy of over-excited explorers, is the flying snake. It is a small, thin creature, exceptionally handsome with green-blue scales flecked with gold and red. Normally, it gives no hint of its aerial prowess. It is, however, a sensationally skilful tree climber, ascending vertical tree trunks at great speed, gripping the bark surface with the edge of broad transverse scales beneath its body and bracing itself by pressing its coils sideways against creepers or any roughness of the bark. Once up in the tree it moves from one to another by racing along a branch and launching itself off. In the air, it flattens its body so that instead of being round, it is broad and ribbon-like. At the same time, it draws its length into a series of S-shaped coils. As a result its body catches more of the air and it glides very much further than it would do if it simply fell. It even seems able, by writhing, to bank and change course in mid-air so that, to some degree at least, it can determine where it will land.

If you now continue to slide down your rope, you will once more come to a layer of leaves. It is not nearly as thick or as dense as the canopy and is formed by the understorey, a few low trees, among them palms, that are specially adapted to the dim light of the jungle floor, and spindly saplings that sprouted from seeds shed by the canopy trees. Once past them you are, at last, back on the ground. As you land, you will feel the forest floor, firm and unyielding beneath your feet, for though it is covered by dead leaves and other scraps of rotting vegetation that have fallen from above, this layer is surprisingly thin. It is very hot and the stagnant air is loaded with moisture. Such conditions suit the processes of decay very well. Bacteria and moulds work unceasingly. Fungi proliferate, spreading their filaments through the leaf litter and erecting their fruiting bodies in the shape of umbrellas and globes, platforms and spikes hung around with lace skirts. The speed of decomposition is extraordinarily rapid. In the chilly forests of the north, a pine needle may take seven years to rot; an oak leaf in a European wood disappears in about a year; but a leaf from a jungle tree decays totally within a mere six weeks of landing on the ground.

The nutrients and minerals released in this way do not, however, remain on the site for long. The daily drenching by rain would soon wash them down into the streams and rivers, so if the trees are not to lose these precious substances, they must reclaim them quickly. They do so with a thick mat of rootlets that they grow close to the surface of the soil. This shallow root system provides little stability for the gigantic trees. Many give themselves additional support with a ring of enormous board-like buttresses that spring from the side of their trunk, 4 or 5 metres above the ground, and extend an equal distance along the ground out from the bole.

This is a dim, twilit world. Less than 5 per cent of the light that falls on the canopy filters down here. That, coupled with the shortage of nutrients in the soil, makes an abundant growth of ground plants virtually impossible. So you will never find a great carpet of colour to rival bluebells in an English woodland in spring. Sometimes you may see a patch of colour ahead of you, but when you get to it you discover that all the

Jungle fungus

flowers are dead. They have fallen from branches up in the canopy. Nonetheless a few flowers do bloom here. Most unexpectedly, to the eye of anyone accustomed only to temperate woodlands, bouquets of blossom sprout straight from the trunk of some trees within a few metres of the ground. The advantages of producing flowers in this way may be connected, indirectly, with the infertility of the soil. If a seed is to grow well here, its parents must provide it with food, since so little can be extracted from the soil. Many trees, therefore, produce nuts loaded with nourishment that will sustain the seedling through the first stages of its growth. Such large objects are more easily borne on the trunk than on the thin twigs at the end of the canopy branches. Down here, too, they are in the clear and easily discovered by pollinating animals. Many of them cater for bats and are pale in colour so that they are easily found at night. The cannonball tree accommodates its visitors even further. It grows a special spike above its flowers from which the bats can hang as they sip the nectar.

One or two flowers grow on the forest floor itself. The plants to which they belong do not, however, draw their sustenance from the soil but suck it from the trees. They are parasites. One of them, Rafflesia, produces the biggest flower in the world. The plant itself exists for most of its life as a mesh of filaments growing entirely within the tissues of a vine root. It only becomes visible when swellings begin to develop along the underground root and eventually erupt above the soil like a line of cabbages. Several species exist in Southeast Asia, but the champion flower is produced by one that grows in Sumatra. The bloom measures a metre across and sits directly on the ground, leafless and monstrous. Its maroon petals, thick, leathery and covered with warts, surround a vast cup, the floor of which bristles with large spikes. From this comes a powerful stench of putrescence. It revolts human nostrils but it attracts flies in swarms, as rotting meat would do. It is they that pollinate it. The seeds, when they develop, are small and hard-coated. No one knows for certain how they are transported to infect another vine, but a likely guess is that they are carried on the feet of large animals that wander through the jungle and may bruise the prostrate stem of a vine, so enabling the germinating Rafflesia to gain entry.

There are not, however, very many such creatures in jungles anywhere, because of the scarcity of leaves on which to feed. In Sumatra, there are a few small forest-dwelling elephants and fewer rhinoceros. They browse on the meagre leaves of the understorey, greatly supplemented by the rich thick vegetation close to the river banks where there is much more light. In Africa a primitive kind of giraffe, the okapi, and in South America the tapir feed in a similar way, but all these creatures are few in number and widely scattered. Nowhere in the jungle can you find large groups of leaf-eaters such as exist in almost all the other terrains of the world. No herds of antelope stampede away from you, no flocks of rabbits look up, panic-stricken, from their nibbling and flee into their holes. The grazers of the jungle are up among the living leaves in the canopy. No large animals can exist by feeding on dead leaves on the forest floor.

But a multitude of small ones can. Beetles of many different kinds, both as grubs and adults, champ their way through decaying twigs and rotting wood. Most numerous and

Buttressed bole of jungle tree, Brazil

widespread of all, termites labour unceasingly in the leaf litter, carrying it away, particle by particle, to their nest. For most of the time they work invisibly within the wood of fallen trees or beneath the surface layer of leaves, but occasionally you will come across a column of them marching twenty or thirty abreast, following a track that has been worn smooth by the infinitesimal patter of hair-thin legs, repeated a million times. They march in a continuous ribbon for hundreds of metres, before finally they disappear into a hole in the ground or a cleft in a tree trunk that leads to their hidden nest.

Cellulose, the material from which the cell walls of plants is made, is very difficult to digest. Dead plant tissue, which has lost its succulent cell contents and sap and consists of little more than cellulose, is very unrewarding material indeed, as far as most creatures are concerned. Some termites deal with it by maintaining colonies of microscopic organisms of a group known as flagellates in the lower part of their gut. These have the ability to break down cellulose and produce sugar from it. The termites not only absorb this by-product of the flagellates' life processes but also digest considerable numbers of the flagellates themselves, thereby obtaining protein as well. Young termites belonging to such species, as soon as they emerge from the egg, equip themselves with cultures of these invaluable protozoans by sucking the rear end of the adults. Many termites, however, use a fungus to help them with their cellulose problem. When foraging workers arrive back in the nest, they take their tiny loads of leaf fragments down to special chambers. Then they chew them up into a kind of spongy compost on which the fungus grows, forming a tissue of interwoven threads. The fungus absorbs nutriment from this compost, leaving behind a honey-coloured crumbly material. It is on this, not the fungus itself, that the termites feed. The young sexually active females, when they eventually fly off to found a new colony, will carry with them spores of this fungus as their indispensable dowry.

Since termites are among the few creatures able to convert rotting vegetation into living tissue, they are a crucial link in the flow of nutrients from one organism to another. Many creatures feed on them. Some kinds of ants exist almost entirely by raiding termite nests and carrying off the grubs and workers for food. Birds and frogs sit beside the marching columns wherever they are exposed, picking off individuals one at a time, while the rest of the column marches doggedly onwards. Pangolins in Africa and Asia and the tamandua in South America, all often called simply ant-eaters, live in fact almost entirely on termites. They have well-muscled forelegs that can excavate termite nests, and long snouts with whip-like tongues that they flick into the wrecked termite galleries to collect the insects by the hundred.

The forest floor can provide a few other vegetables apart from dead leaves. Nuts and fruit, fallen from above, are easily collected and tubers and roots can be dug up. There are even a few buds and leaves from the shrubs of the understorey. In each jungle continent at least one species of mammal finds enough of these things to live on – in Asia the mouse deer, in Africa the royal antelope, and in South America the agouti. These three belong to very different families. The mouse deer is related to pigs and primitive ruminants; the royal antelope is indeed a true antelope though an exceedingly

Rafflesia

small one; and the agouti is a rodent. Yet they all look very similar – about the size of a hare with pencil-thin fragile-looking legs that end in sharp claws or hooves so that the animals appear to be running on tiptoe. And they all have very similar habits and temperament, being extremely nervous, freezing when alarmed and then sprinting away in frantic zigzags across the forest floor. The mouse deer and the agouti even use the same method of signalling among themselves. They do so with tiny, impatient stamps. And they all feed on leaves and buds, fruits, seeds, nuts and fungi.

Many birds also find enough to sustain them on the ground and seldom leave it, only fluttering into the branches under extreme provocation. One such is the jungle fowl, the ancestor of our farmyard chickens. It is still common in Malaysia, producing in the early morning a high pitched and slightly strangulated version of the familiar cock-crow that, illogically, sounds incongruous in the tropical jungle. Curassows, black turkey-like birds, are their equivalent in South America. One or two of these ground birds have grown so big that they can hardly fly at all. The argus pheasant of Southeast Asia is one of the most spectacular. The female is also rather like a turkey in shape and size, but the male is very different indeed. He grows a huge tail, a metre or so long, and immense wing feathers, each decorated with a line of large spots that look extraordinarily like eyes. It is these that caused the bird to be named after Argus, the many-eyed monster of Greek myth. The male clears an arena in the forest, 6 metres or so across, and keeps it clean and free of fallen leaves and twigs. He will even peck around the base of a seedling tree to kill it, if he cannot uproot it. He summons a female to this ground by loud calls that echo daily through the forest. When she arrives, he leads her to his arena and begins to dance before her, becoming more and more excited. Suddenly he erects his enormous tail and fans his wings so that he is transformed into a towering screen of feathers, studded with line upon line of lustrous simulated eyes.

Several birds of paradise in New Guinea also maintain dancing arenas on the ground and display in a similar way. The six-wired bird of paradise dances by standing upright, spreading a velvety-black skirt of feathers and nodding the six pennant feathers on its head. The superb bird of paradise performs on a low branch with a huge triangular shield of iridescent green feathers on its breast, which flash in the dim light. In South America, the great dancer is the cock-of-the-rock. It performs, not singly on isolated stages, but in groups of a dozen or so. The male is a marvellous orange with black wing feathers and an orange semicircular crest that comes right down over the front of the head and almost conceals his beak. During the breeding season, they assemble at one place in the forest. Each male claims a small arena of his own on the ground. Most of the time the birds sit in the branches of saplings or lianas beside the arenas, but when the drab brown female appears, they all flop down on to their own arenas with a squawk and begin their displays. They crouch with heads cocked to one side so that their crest is horizontal. They bounce up and down, snapping their beaks with audible clicks. Sometimes they freeze, motionless but tense. Eventually, the female flutters down to one of the arenas and nibbles the outer filamentous feathers of the owner's rump. Quickly he hops up and mounts her, there on the dancing floor. The coupling lasts only

Tamandua ant-eater

a few seconds. Then the female flies off into the forest where, unaided, she will lay her eggs and raise her young, inconspicuous in her brown feathers, while the male, bright as a flame, continues to bounce and bob on the forest floor.

The most wide-ranging and omnivorous inhabitant of the jungle floor is, of course, man. He first evolved in the open savannahs, but it is likely that he invaded the jungle at quite an early stage in his history. To begin with, doubtless, he was a wandering hunter as the Zaire pygmies, some of the orang asli of Malaysia and some Amazonian Indians still are today. These people are all of small stature. Indeed, the Mbuti of Zaire are the shortest of all human beings, their men having an average height of less than 1.5 metres and their women being even smaller. The relative poorness of their diet may have something to do with this, but it is also undoubtedly true that their small size suits them very well for life in the jungle, allowing them to move swiftly and quietly through the forest. Their bodies are slim and largely hairless, and they sweat very little, for this method of cooling the body, effective though it is in other parts of the world, does not work well in the jungle where the air is so humid that moisture on the skin evaporates only slowly. Travellers from cooler parts know this only too well. Their sweat cascades from their skin, soaking their clothes, but does nothing to lower their temperatures. Their guides, meanwhile remain dry-skinned, cool and untroubled.

These nomadic people understand the jungle intimately and in detail. They know better than any other creature how to extract food from every part of it. They collect tubers and nuts from the floor. They cut open fallen trunks to pick out the edible beetle grubs, climb trees to pluck fruit, pull honey-laden combs from the nest of wild bees and cut those particular lianas which, for a few moments, will spout water like a tap and give them a drink. And they are skilful and brave hunters. The Mbuti catch royal antelope and okapi in nets and go on long and dangerous hunts to kill forest elephant. All know how to imitate the calls of ground-dwelling birds and mammals and summon them within range of spear and arrow. Since the great majority of animals live in the canopy, the people have had to develop weapons with considerable range. The South American Indians use blowpipes. An inner section of thin bamboo or a tall reed, cleared of its sectional walls, is put inside a wooden outer case that gives it protection and rigidity. The darts, poisoned at one end and fitted with fluffy seed fibres around the other so that they make an airtight fit in the pipe, can be blown with such force that they easily reach a target 30 metres above. Their poison may be so potent that a well-struck animal will collapse and lose its grip within a minute or so; and their discharge and flight so soundless that even when one bird in a flock is hit and falls, the rest may well stay unalarmed, allowing the hunter to claim another.

The nomads, like all people everywhere, do not live by food alone and the forest provides them with much else besides. Frogs, roasted on spits, exude poison for the blowpipe darts; fibres from vines make nets; resin exuding from certain trees makes excellent torches; palm leaves will serve as watertight roofs for shelters. When festivals and rituals are held, crushed seeds will produce a paint with which to decorate the body; and parrot feathers and hummingbird skins furnish head-dresses of the greatest splendour.

Superb bird of paradise, New Guinea

The nomadic life is nonetheless a hard one and the search for food time-consuming and arduous. Many jungle dwellers prefer to hack clearings in the forest and make gardens. Originally, they used axes armed with small stone blades, laboriously chipped and polished. Even with metal blades, the work of felling trees is long and hard. After the forest has been cut down and the foliage and branches burnt, cassava or corn, taro or rice is planted among the prostrate trunks. But the poverty of the soil is such that after three or four seasons, crops will no longer give a worthwhile yield and the people must move on and clear another plot.

Forest trees fall eventually, whether or not man cuts them down. Many stand for several centuries, but eventually the sap no longer rises so strongly in the giant trunks. The aged branches, attacked by moulds and fungus, riddled by tunnelling insects, can no longer carry their burden of leaves and squatting plants. One big branch breaks and the tree may be fatally unbalanced. The end is likely to come during a storm. The torrential rain adds several tons to the weight of the lop-sided crown. A strike of lightning will deal a final blow. Slowly the immense tree hinges over. The lianas binding it to its neighbours tighten. Some snap. Others drag at the surrounding branches. As the crown topples forward with increasing speed, it tears through the canopy with a roar of sustained smashing. As the first branches hit the ground, a fusillade of cracks, like rifle shots, ring out, followed seconds later by the earth-shaking thundering double thump as the immense trunk hits the ground and bounces. Then there is silence, broken only by the soft patter of leaves torn from the branches by the rush of air, gently raining down over the wreckage.

The death and fall of an old tree, that has destroyed the homes of birds and snakes, monkeys and frogs, has also brought the promise of life to the small saplings that till now stood in its shade. Many have remained only a foot or so high for ten years or so, waiting for this moment. For them, a race has now begun. The prize and the finishing post is the gap torn in the jungle canopy by the fall, through which the sun now shines. The strong, unaccustomed light, the first they have experienced in their lives, triggers their growth. Fast though they now put out leaves and branches, and grow upwards, they are outpaced by others. Seeds that have lain dormant in the ground rapidly sprout. Banana palms and ginger plants, heliconias and cecropias, all plants that live in the sunshine of river banks or in forest clearings, quickly spring to life and put out big broad leaves to soak up the sun, to flower and to fruit. But within a few years, the pack of saplings are once more above them. As they grow, one or two, because of their natural vigour, a propitious start, or a rather more nutritive patch of soil, take the lead. As they spread their branches, they overshadow their competitors. Deprived of sunlight, the lesser trees become feebler, drop out of the race and die. After several decades just one or two attain their full height and can begin to flower. The canopy of the jungle once more has been closed and the stability of the life beneath restored.

The jungle, Ecuador

FIVE

SEAS OF GRASS

As you travel through forests, whether tropical jungle or temperate woodland, away from permanent open water and towards a drier part of the country, the trees will begin to dwindle in number and stature. Bulky trunks, branches and leaves require a minimum amount of water if they are to be sustained. So if the rainfall is low, or if the soil is so sandy and well-drained that even at depth it lacks moisture, then trees will not grow, the forest will come to an end and you will emerge on to open plains covered with grass.

The name, grass, covers a multiplicity of plants. Indeed, the grass family is one of the largest in the plant kingdom containing, worldwide, about 10,000 different species. Grasses are not, as you might suppose from the simple character of their leaves, primitive plants, but highly evolved ones. Their flowers are often not recognised as such. Grasses, growing as they do in open, treeless country where there is nearly always a breeze, can rely on the wind to distribute their pollen. Since they do not need to attract animal pollinators, their flowers have no need to be conspicuous or brightly coloured. Instead, they are small and drab, with tiny scales instead of petals, and grow in clusters on special tall stems that lift them into the path of the wind.

The one condition that grasses require is good light. They cannot grow in the deep shade of the forest. But they can tolerate many other hardships that would cripple or kill other kinds of plants. They can withstand not only low rainfall but also scorching sunshine. They can survive fire, for even though the flames sweeping over them may burn the leaves, the root stock lying close to the surface of the soil is seldom damaged. They can even tolerate regular mutilation by grazing teeth or the blades of a lawn-mower.

This remarkable endurance comes from the particular way in which grasses grow. The leaves of most other plants spring from buds on a stem, develop a branching network of veins to carry their sap, and quickly expand to their final shape. Their growth then stops. If they are damaged, they can seal off their broken veins to prevent the leakage of sap, but they cannot repair themselves further. The leaf of a grass is different. Its veins form, not a network, but a row of straight unbranched lines that run up the length of the leaf. The growing point is at the base of the leaf and it remains

Fox-tail grass

active throughout the life of the plant. If the upper section of the leaf is damaged or cropped, then it grows at the base to restore its original length. Furthermore, the grass plant itself spreads not only by means of seeds, but by putting out horizontal stems along the surface of the ground, and each joint of these is able to sprout leaves and roots.

The roots of grass plants are fibrous and grow so profusely that they create a matted tangle that extends for several centimetres below the surface. This turf holds the soil together even during a prolonged drought, preventing it from blowing away, and when rain does eventually fall, green leaves can be produced within a day or so.

These efficient, persistent plants evolved comparatively recently. They were not in existence when the dinosaurs were alive, so those creatures had to survive on the rather coarser fare of ferns, cycads and conifers. When new kinds of trees in the forests began producing the first flowers, and lilies in the lakes sent up their blossoms to star the waters, the dry flat lands beyond the woodlands were still bare earth. Only long after the age of reptiles, some 25 million years ago, when the mammals were launched on their great expansion, did grasses begin to colonise the plains.

Today, grass plants cover about a quarter of the surface of the land. Each country has given its grassland its own name – the pampas and the campo in the southern part of South America, the llanos on the plains around the Orinoco in the north; the prairie in North America and the steppes in central Asia; the veldt in the south of Africa and the savannahs in the east. These are areas of great fertility. Individual grass plants may live only a few years before being replaced by new seedlings. Their dead leaves build up a mat of decaying vegetable matter that lightens and enriches the soil beneath, making it crumbly and well-aerated. Among the grass plants and, to some extent, shaded and protected by them, grow many small flowering plants – vetches which fix nitrogen in nodules on their roots, daisies and dandelions with flowers made up of a mass of tiny florets, and plants from other families which store food in bulbs and swollen roots. The eternally springing grass, surviving drought and flood, grazing and burning, is lush and sappy in moister regions, dry and tough but nonetheless edible in drier parts, and presents an easily-gathered banquet for a multitude of animals. Indeed, a hectare of grassland is capable of supporting a greater weight of living flesh than any other kind of country.

The miniature jungle created by the tangled roots, matted stems and clumps of growing leaves has its own community of tiny inhabitants. Grasshoppers chew away at the living leaves; aphids and bugs pierce the veins with their needle-shaped mouthparts and suck the sap; beetles munch the dead leaves. In temperate regions earthworms writhe out of their burrows to collect dead leaves and carry them down for digestion below ground; and throughout the tropical grasslands termites labour.

The skin of a termite, being soft and thin, does not retain moisture very effectively. In the humid air of the jungle, this caused no problem and columns of workers marched openly across the ground; but such behaviour would be almost lethal out on the open plains, for the sunshine would dry out the tiny bodies, and they would shrivel and die.

Termite mound

One or two species do manage to travel unprotected above ground during the cool of the night, but most grassland termites travel by tube, tunnelling beneath the surface of the soil or roofing over their trackways with ceilings of masticated mud. When such species set about demolishing a small bush, they will first enclose the entire plant in a wrapping of bulging mud walls and then work away tirelessly in their dark humidity.

The need to conserve moisture also dictates that a termite colony must build a nest for itself. Some species excavate their chambers and galleries below ground. Many build enormous mud fortresses. Each labouring insect makes its own bricks by chewing earth, mixing it with liquid cement from a special gland above its jaws, and producing a small pellet which it kneads into position on the rising wall with a shaking action of its head. Millions co-operate to construct their immense tenements. These may measure as much as 3 or 4 metres across. Some may have lanky spires 7 metres high. Ventilator chimneys run up within the buttresses around the sides to allow spent air to escape. Deep shafts descend through the foundations to moist ground where the workers go to collect water. This they smear over the internal walls of their galleries and so prevent a lethal drop in the humidity of their micro-climate.

Ants, too, live on the grasslands. Superficially, they may resemble termites, but they are very different insects. Whereas termites belong to the same group as cockroaches, ants are related to wasps, as can be seen from the wasp-waist that they possess and termites lack. Ants, also like wasps and unlike termites, have a hard, impermeable outer skin to their bodies, so they are able to march above ground, even in sunlight, with little risk of desiccation. Harvester ants swarm through the turf, indefatigably collecting grass seeds and carrying them back to their underground granaries. There, workers belonging to a special caste with huge jaws crack them open so that other less well-equipped members of the colony can eat them. Other species, the leaf-cutters, demolish living plants, using their scissor-like jaws to shear the leaves and stems of the grasses into easily transported sections.

Ants cannot digest cellulose any better than termites. They too recruit the help of a fungus. It is not the same species as that cultivated by termites, and the ants eat it directly. The nests of the leaf-cutters are not as obvious as termite hills, for they are built below ground; but they are even bigger. The galleries may go down to a depth of 6 metres, extend over an area of 200 square metres and provide a home for seven million insects.

Other kinds of ants tap the nutriment of grass by using as an intermediary, not fungi, but aphids. These insects digest only a small part of the sap they suck. The rest they excrete as a sugary liquid that is known, somewhat flatteringly, as honey-dew. It can often be found as a sticky film on the ground beneath an aphid-infested plant in the garden. Some ants, however, find honey-dew an excellent food and they herd the aphids into flocks and milk them in much the same way as human farmers tend their herds of dairy cattle. The ants encourage the aphids to produce more honey-dew than they would normally do by stroking them repeatedly with their antennae. They protect them by driving away any other insect invading the aphids' grazing grounds with

Leaf-cutting ants

squirted volleys of formic acid. Some build special shelters of parchment or earth around a particularly productive stem or root on which the aphids are grazing and so deprive them of their free-range, like animals in a factory farm. At the end of the summer, when the aphids die, the ant-farmers take the aphid eggs down to their nests for safe-keeping. When young aphids hatch out in the following spring, the ants carry them out again and put them to graze on fresh pastures.

All these insects – aphids and ants, termites and grasshoppers, bugs and beetles – are themselves potential food for other bigger creatures. The grasslands of South America are patrolled by one of the most unlikely-looking of all living mammals, a creature to match the most extravagant heraldic design. It is the size of a large dog. Its head is elongated into a long curving probe, with eyes and tiny ears at the top end, and nostrils and a small narrow mouth at the bottom. Its body is covered with bristly hair and its gigantic tail, which makes up half its length, is so extravagantly tufted along its upper and lower surfaces that it sails out behind the animal like a banner.

This is the giant ant-eater. Its eyesight is very poor, its hearing scarcely more acute, but its sense of smell is excellent and it is able to locate termites by the scent of their dried saliva mixed in the walls of their mounds. Once the nest is discovered, the ant-eater widens the entrance of one of the main tunnels with the long curved claw on its foreleg and inserts its snout. From the end of this comes a long thong-like tongue which whisks down the termite corridors at great speed, sometimes as frequently as 160 times a minute. Each time it flicks out, it carries a fresh coat of saliva, and each time it is withdrawn it brings with it a load of termites. These are scraped off inside the tunnel of its toothless mouth and swallowed whole. Its stomach is very muscular and contains small quantities of gravel and sand which help to crush the insects as they are churned about and finally digested. In such a way, an adult giant ant-eater can consume, in a single day, some 30,000 termites.

Other less specialised feeders, the armadillos, also take their share of ants and termites. As their name suggests, armadillos are armoured, with a flexible horny shield backed by bone over the shoulders, another over the hips and a varying number of bands around the waist connecting the two. The most dedicated termite-eater among them is the giant armadillo. It is comparable in size with the giant ant-eater, but it has a much more full-blooded attitude to collecting its food. Instead of fastidiously poking an elegant nose into an exit corridor, the giant armadillo excavates a huge tunnel, hunching its armoured back against the roof, hurling out earth with sweeps of its forelegs, until finally it reaches the heart of the termite colony, oblivious, it seems, of the bites of thousands of angry termite soldiers.

Its smaller armadillo relations, the seven-banded, the hairy and the spiny, are all more broadminded in their dietary tastes and take not only ants and termites, but nestling birds, grasshoppers and even fruits and roots. The three-banded is the only one which is able to protect itself by rolling up. As it snaps shut, the triangular shield on the base of its tail fits alongside the triangular shield on its head so that its whole body forms an impregnable armoured ball the size of a grapefruit. Its bigger relations also

Giant ant-eater

have little to fear from predators such as foxes or hawks. The giant armadillo, like the giant ant-eater, is altogether too big and can, in any case, deliver fearsome blows with its digging forelegs, and the smaller ones are sufficiently well-armoured to repel initial attacks and can thwart any more sustained assault by digging their way down to safety.

The grazing insects do not, of course, have the grass to themselves. Little brown cavies, snub-nosed and tail-less, the wild ancestors of domestic guinea pigs, make tunnels through the grass along which they patter back and forth, harvesting the juicy stems. Viscachas, larger rodents the size of a portly spaniel, live in underground warrens and emerge in the evenings to crop in a leisurely way the turf within easy reach of their burrow entrance, so that if there is a hint of danger, they can easily run back to safety. The mara, another cavy and a bit bigger still, forages more widely and during the day. Away from its hole, it relies for its safety on its speed. It has long slender legs and the highly-strung skittish disposition of a European hare, with a tendency to execute huge soaring leaps at the most unexpected moments.

These grass-eaters are hunted by many predators. Caracara hawks stalk across the grass and pounce on the cavies. So does the pampas fox, a creature similar in appearance to a jackal. An even bigger member of the dog family, the maned wolf, also roams the pampas. It looks more like a fox than a wolf, but one that is being viewed in a fairground distorting mirror, for although its head is little more than the size of that of a sheep dog, its legs are so long that it stands a metre high. These long limbs may enable it to run at great speed, though it is difficult to know why it should need to do so. There is no report of it being pursued by any larger animal – except man – and speed is not needed to catch cavies. Indeed, its taste is for smaller rather than larger prey such as nestlings, lizards, even grasshoppers and snails, and it also eats roots and fruit.

The biggest of the animals on the pampas is not a hunter at all but a grass-feeder. It weighs more than the maned wolf and stands twice as high. It is not, as you might guess, another mammal but a bird, the rhea. It looks like an ostrich, being flightless with fluffy useless wings, a tall neck and long bony legs that give it great speed. Although it eats a wide variety of things, including insects and small rodents, its main food is grass, and at some times of the year, rheas form flocks on the pampas comparable to grazing groups of antelope.

A rhea's nest, if you are unprepared for it, is a quite extraordinary sight. Each egg is ten times the size of a chicken's egg. That might be expected from such a large bird, but a rhea's nest often contains at least twenty of them and cases have been recorded in which there have been over eighty. They do not, however, all come from a single female. The male is polygamous. He builds the nest, such as it is, by clearing a shallow depression on the ground, usually in a patch of scrub or tall grass, and lining it with dry leaves. He courts and mates with a number of females, dancing round each of them, with swaying neck and ruffed feathers. As the pair become more and more excited, they may even entwine necks. Then the female squats and he mounts her. Soon after, the female will visit him as he sits on the nest and he will rise to allow her to deposit her egg in it. Female after female comes to him. If one finds the nest occupied by another,

Rheas

Burrowing owls

she will lay her egg outside the bowl and leave it to the male to roll it in with his beak to join the main clutch. Sometimes so many females contribute so many times that when he starts to incubate, there will be more eggs than he can possibly cover. So he will abandon the surplus outside the nest, where they go cold and addle.

A sitting male rhea is a formidable guardian. Any other creature venturing near the nest is likely to be charged and driven away. So the rhea has no need to make its nest inaccessible. But none of the other birds on the pampas has the rhea's size and strength and for them finding a secure nest site is a major problem.

The oven bird is one of the few that manages to construct a near-secure thief-proof nest entirely by itself. It will build on a post or the low branch of an isolated tree. The material it uses is simply mud mixed with a little grass, but with it the bird constructs a rock-hard, domed chamber with a partition wall just inside the entrance hole that makes it almost impossible for a snout or a paw to reach the eggs or nestlings beyond.

The flicker, a kind of woodpecker which here on the plains eats mostly ants, often uses termite hills for its nest site. It has retained enough of its ancestral skills to be able to excavate a hole in the termites' hard masonry. The termites themselves then repair the broken galleries inside, stopping them off so that the flicker is eventually provided with a smooth-walled chamber for its eggs.

Holes in the ground, dug originally by foraging armadillos or inhabited by viscachas, are often commandeered by tiny owls. They are quite capable of digging for themselves and indeed sometimes do so, but they seem to prefer to be lodgers. Often every hole in a viscacha warren will have its owl standing like a sentinel at the entrance. As you approach, it glares at you with piercing yellow eyes, bobbing up and down in comic agitation until, right at the last moment, it loses courage and ducks down into the security of its borrowed burrow.

The caracara hawk prefers a small tree if it can find one, but if necessary it will nest on the ground. With its powerful butcher's beak and talons, it is sufficiently well-armed to drive off most creatures and will eat lizards and snakes. The spur-winged plover, however, is a much smaller bird and, being a general feeder on insects and other small invertebrates, has only a small beak and no large claws. It seems to have little defence against its eggs being eaten by reptiles or prowling armadillos, but the plover is a valiant protector of its nest as you will soon discover if you happen to walk near. It will swoop down from the sky with a rush of wings and piercing screams. It may even strike your head with its wings. If this fails to deter you, the bird will land, spread one wing as though crippled and continue to shriek loudly. It is commonly said that it is feigning injury. Certainly the performance is sufficiently extraordinary to make you, and presumably other creatures, walk towards it to investigate and so move away from its nest. Sometimes it is even more devious. It lands, settles down in the grass with half-open wings and begins to gather little pieces of grass around it with its beak, exactly as though it were sitting on its nest. If you go over to investigate the exact spot, it will take off again and only then will you discover that you have been fooled and there is nothing there. If all this fails, the plover has one further defence. The eggs and chicks are so perfectly camouflaged that even when you step within inches of them, you may not notice them. This combination of strategies seems to be very effective, for in some parts of the grasslands, spur-winged plovers are everywhere and their cry of *tero-tero* the most common and evocative sound on the pampas.

The very flatness and uniformity of the plains, the unvarying character of their grass mantle, produces animal communities that, compared with those inhabiting jungles and woodlands, contain relatively few species and are simple in their inter-relationships. The grass is grazed by insects and rodents. Bigger grazers produce dung, which lies on the plain and is reworked by insects or washed by rain back into the ground. The insects are eaten by armadillos and birds; the rodents by hawks and flesh-eating mammals. When the hunters die, the substance of their flesh is returned to the soil either by way of scavengers or directly by the processes of decay. So the nutrients synthesised by the grass are returned to it, thus ensuring that it will continue to sprout and feed fresh generations of grazers.

Such communities, with minor variations, extend from the cool pampas of southern Argentina north for 3000 kilometres across the campo around the River Plate, up into Paraguay and beyond into southern Brazil. On the southern edge of the Amazon basin, however, there is sufficient rain to enable trees to grow. That is the end of the grasslands and the beginning of the jungle.

One thousand five hundred kilometres farther north still, on the other side of the Amazon, around the middle stretches of the Orinoco River, stretch more grasslands that are known as the llanos. If you go there in December, you will see a landscape very reminiscent of the pampas. Acres of grass ripple in the wind beneath high clouds in a blue sky. Yet the animal population is strikingly different. Some of the birds – spur-winged plovers, caracara hawks – are, indeed, the same, but there are no cavies in the grass, and no viscacha warrens. Stay on the llanos for a few months more and the reason for their absence – and for the lack of trees – becomes obvious. Storm clouds gather. The sky becomes overcast and rain begins to fall in torrents. The rivers rise at an alarming speed, fed further by storms on the flanks of the Andes 500 kilometres away to the west. Eventually they overflow. The soil here is thick clay. The water does not drain but spreads as a shallow flood across the llanos. The roots of any tree growing here would be water-logged, and any burrowing animal would certainly be drowned.

Now the major grass-feeder of the llanos comes into its own. The capybara is the biggest of all rodents. It is the size of a domesticated pig and has been called the Orinoco hog. Its coat is long and brown and it has webbed feet to assist it in swimming. Its eyes, ears and nostrils are all on the top of its head so that it can lie in the water almost completely submerged and yet be aware of what is going on around it. It lives in rivers, lakes and swamps all the way from Argentina to Colombia, in both grassland and jungle, grazing on water plants or on grass and other vegetation close by the banks. When the floods come to the llanos, its range suddenly expands from a thin strip alongside the river into vast lagoons. The capybara takes full advantage of this new freedom. Family groups, twenty or thirty strong, splash through the shallows, pulling

Capybara

up the drowned grass, and swim in squadrons across the deeper stretches. No other grass-feeder here, whether mammal or bird or insect, is as amphibious as they are, and for a few months they have all this great pasture to themselves.

North and west of the llanos, in Panama, Guatemala and southern Mexico, the jungle returns, but beyond, across the border of the United States, on the prairie of southern Texas, the grasslands reappear. The American prairie stretches in a band some 3000 kilometres long and 1000 wide up along the eastern side of the Rockies through Oklahoma and Kansas, Wyoming and Montana to the Canadian border and beyond, right to the southern edge of the northern forests. It is the greatest and the richest of all the grasslands in the world.

There are few termites here and no specialised ant-eaters, but otherwise there are equivalents to most of the creatures of the pampas. The grass is alive with insects. They in turn are the food of a multitude of birds. The burrowing colonial viscachas are paralleled by burrowing colonial prairie dogs; the pampas fox by the coyote; the caracara by the red-tailed hawk. But in one respect the prairie differs in a most spectacular way. The largest herbivores wandering across the plains are not birds like rheas, but huge mammals, bison.

Bison are wild cattle. They belong, together with antelope and deer, to a huge family of mammals that have evolved a special way of digesting grass. They ruminate.

The stomach of a ruminant is divided into special compartments. The first chamber, which receives the mouthfuls of half-chewed grass when they are initially swallowed, is the rumen. It contains a rich broth of bacteria and protozoans which set to work, breaking down the cellulose in the leaves, just as similar micro-organisms do on a tiny scale in the gut of some termites. After several hours, the mash of half-digested leaves is separated into lumps by a muscular pouch alongside the rumen and these are sent back up the throat one at a time, to be given a second long grinding by the teeth – the process known as chewing the cud. When eventually the lump is swallowed for the second time, it by-passes the first two chambers and arrives in the third, the stomach proper, where, after being worked on further by more digestive juices, the nutriment from the grass is finally absorbed through the stomach wall.

Ruminating animals evolved somewhere in the northern continents some 20 million years ago and spread very widely, west into Europe, south into Africa, and east into North America. Their colonisation of South America, however, was very erratic, for the land bridge, formed by the isthmus of Panama, has not always existed: for long periods, South America was an island, cut off from the rest of the world. Their only representatives here are a few species of deer and the llamas. In North America, however, the ruminants flourished, and when Europeans first reached the prairies, they found on them herds of a size that beggared both belief and description.

A bison bull is a huge creature, the largest and heaviest of all living animals in America, standing nearly 2 metres high at the shoulder and weighing 1000 kilos. Travellers across the prairies only 150 years ago called them, inaccurately, buffalo, and described herds that covered the land with a rippling flood of brown backs, stretching

Bison, Yellowstone

to the horizon in all directions. One wrote of a dense herd travelling at a steady gallop that took over an hour to pass him. Several attempts have been made to estimate how many bison there were on the prairies at this time. Even the most cautious put the figure at around 30 million and some authorities believe that there were twice as many.

The bison fed on the grass plains of the northern part of their range during the summer. In autumn, when the growth of the grass ceased, they migrated south for 500 kilometres or more along tracks that were so well established and so deeply trodden that the human settlers themselves adopted them for their own journeys.

Living with the herds were tribes of Plains Indians. They hunted the buffalo with bows and arrows and obtained from them almost everything they needed. The flesh they ate. The skins they used for clothing. The horns were carved into cups, the bones into tools. Ropes, bags, sledges, tent-coverings, all came from the bison. The bison also provided the Indians with the image and the spirit of their gods. No people have ever been more intimately involved with an animal species.

Though the Indians exploited the buffalo so thoroughly, they took only what they wanted for their own immediate use. Not so the first white settlers. The bison consumed grass that could be turned into more saleable flesh – beef. Their trampling hooves prevented the replacement of the native prairie grasses with domesticated strains which produced flour – wheat. In any case, getting rid of the bison was an indirect way of getting rid of the unwanted Indian who, without them, would have difficulty in surviving. So the bison had to be destroyed.

The massacres started around 1830. The settlers were not now killing for food. They were shooting the animals just to be rid of them. In 1865, the railway was built across the continent from east to west, cutting the bison population into two. No longer could the northern herds migrate southwards unimpeded. The famous hunter, Buffalo Bill Cody, was employed by the railway to provide meat for the construction gangs. He alone killed over 4000 bison in eighteen months. Passengers on the railway were encouraged to shoot the great beasts from a moving train as sport. Sometimes they cut the tongue from a dead animal. It was a delicacy. For a short time, luxurious travelling robes made of buffalo hides were fashionable so skinning the animals was worthwhile. But for the most part the mountainous corpses were simply left to rot.

For several years in the early 1870s, two and a half million were killed annually. By the end of the decade, the bison had been exterminated south of the railway. In 1883, a herd of 10,000 in the north was obliterated in only a few days by the simple strategy of stationing gunmen at every known waterhole within the herd's range. All the animals had to drink. All the animals were shot.

By the end of the century, there were less than a thousand wild bison left in the whole of North America. And then, at this very last moment, action was taken to protect them. A group of naturalists, with Government help, managed to assemble the survivors and unite them with others that had been kept in zoos and private parks. Slowly the numbers were built up again. Today, there are about 35,000 bison living on tracts of prairie that are conserved as national parks. But that is as many as there is ever likely to be in the future, no matter how carefully they are protected. Men are not likely to spare any more land for them.

Sharing the prairie with the bison were herds of another ruminant, an antelope-like creature called, because they have short two-branched horns, pronghorn. They are neither true antelope nor true deer but a primitive intermediate group. Once they rivalled the bison in number. Estimates of their population during the nineteenth century vary between 50 and 100 million. Lacking the huge bulk and power of the bison, they were more vulnerable to predators such as the wolf and they relied for their defence on speed. They are the fastest wild animal in North America, being able to reach speeds of 80 kilometres an hour. But that did not save them from human hunters. They were also killed without mercy and by 1908 only 19,000 were left. Fortunately, they too are now protected and their numbers today are approaching half a million.

The ranges that once supported the vast herds of pronghorn and bison are now grazed by imported strains of domesticated cattle. Clearly people want meat with which to feed themselves. But, ironically, the prairie grass can support only a third of the weight of cattle that it can of the creatures that originally evolved to exploit it.

The grasslands of central Asia lie in much the same latitudes as the American prairies. They are not, for the most part, nearly so fertile for, placed as they are in the centre of the greatest land-mass on the globe, they receive very little rain indeed. The soil over great areas is dry and dusty in the summer and deeply frozen for much of the winter. Nonetheless, great herds of ruminants live here too. Saiga antelope are true members of the antelope family but very odd ones. They are the size and general shape of sheep, but they have the most extraordinary heads. The eyes are huge and bulbous; the horns, carried by the male alone, are amber coloured and simple upright spikes; and, oddest of all, the nose ends in a stubby flexible trunk. The nostril openings are wide and circular. Inside they have a convoluted arrangement of glands, mucous tracts and sacs which take up so much space that the front of the animal's head bulges out, giving a Roman nose to its muzzle. The function of this extraordinary apparatus is to warm and moisten the air and filter the dust from it.

The animals move continuously across the steppes cropping the meagre grass. They are able to sense impending changes in the weather, for they will suddenly alter their gait from a slow amble to a smart trot and travel at speed for several days to escape an impending blizzard.

During the eighteenth century, they were widespread from the shores of the Caspian Sea in the west to the edge of the Gobi Desert in the east, and so abundant that tens of thousands were often killed in a single hunt. As more men with better firearms began to travel over the steppes, the saiga were hunted with increasing intensity, for their meat was greatly relished. By 1829 they had been exterminated over the whole of the middle of their range between the Urals and the Volga, and by the beginning of this century they were reduced to under a thousand individuals. It seemed that the animal must inevitably become extinct.

Then someone realised that no other animal of any kind, wild or domestic, could

convert the grass of the steppes into meat with such efficiency as the saiga. If they disappeared, great areas of the steppes would produce nothing whatever to feed mankind. So hunting them was prohibited and the survivors protected and managed with as much care as if they had been pedigree cattle.

Their recovery was extraordinary. It seems that the animal is, by nature, specially adapted to respond to a great reduction of its numbers caused by natural disaster such as extreme drought in the summer or a winter of great severity, for the females have a quite exceptional breeding rate. They mate when they are only four months old and not yet fully grown. While they are pregnant, the young females grow very little, but once the calves are born, they begin to grow again. By the beginning of the next breeding season, they are full-sized. Thereafter, three-quarters of the females will give birth to twins. As a result of this extraordinary fecundity, the saiga were able to recover swiftly from the biggest of all disasters that had ever overtaken them, their first contact with men carrying guns. In fifty years they have increased from a few hundred to over two million. So today, the Soviet Union crops a quarter of a million of them every year for meat.

The same story of mass herds and mass slaughter was played out on the veldt of southern Africa, but here, for one species at least, the last moment reprieve never came. As European colonists, settled around the Cape, began to push northwards in the early nineteenth century, they discovered rolling grass plains thronged with vast herds of antelope of several different kinds – springbok and blesbok, hartebeeste and white-tailed gnu. The springbok were so numerous that they made regular mass migrations in search of new pastures. On these occasions, they formed herds so big that the whole landscape appeared to be on the move. Their numbers dwarfed even those of the pronghorn and bison in North America. One naturalist in 1880 believed that just one herd on such a migration contained at least a million springbok.

There were also great numbers of another kind of large grass-feeder, one that has played a major role in mankind's history – the horse. Ancestral horses had evolved originally on the grasslands of North America. They too employed bacteria and protozoans in their stomachs to help them digest the leaves, but managed to do without the complex gastric apparatus of the ruminants. For a long period they were very successful and they spread across the Bering Strait land bridge, which then existed, into Asia, Europe and down to Africa. In America they eventually yielded their place to the early cattle and antelope, and disappeared. In Europe and Asia, they and their near-relatives, the wild donkeys, were first hunted and then domesticated. Today the wild forms there are nearly extinct and survive only as a few small herds in Central Asia. Only in Africa do large herds of them still gallop. And magnificent creatures they are, handsomely striped in black and white. Several kinds were eventually discovered – a narrow-striped one, Grevy's zebra, that lives in parched country close to the edge of the Sahara; two kinds of mountain zebra in the west; and on the veldt, five kinds of plains zebra. One of these, the quagga, was not completely striped. It had such markings on its head and neck, but its body was plain brown, fading to white on the legs.

Pronghorn antelope

All these animals, antelope and zebra, were classified by the settlers as game – creatures that could be hunted for food or for fun. By 1850, hunters began to notice that the game was not as abundant as it had been. The killing, however, went on unabated. The virtual destruction of the herds took a mere thirty years. By the end of the century, the blesbok herds had been reduced to about 2000 individuals. Springbok existed only in tiny isolated groups. There were less than a hundred mountain zebra. No white-tailed gnu at all were left in the wild and only about 500 survived as captives on farms. And the quagga had been exterminated. It had not been regarded as particularly good eating. Its hide was valued more than its flesh, for it could be made into shoes and light serviceable sacks. But it was easily found and easily shot. The last wild one was killed in 1878. The last lonely captive died five years later in a zoo.

The only one of the great grasslands of the world that still retains its populations of large grazing animals almost intact are the savannahs of eastern Africa. The survival of the herds there is due largely to the fact that the land is not as well-watered as the prairie, the veldt or the pampas and therefore not suitable either for man's domesticated creatures, which are all descended from temperate species, or for growing his domesticated grasses. These lands today support the largest concentrations of big wild mammals in existence.

The savannah country forms a vast horseshoe around the West African jungle about a million square kilometres in extent. It is much more varied in character than the other great grasslands. Low thorn bushes are common in many parts of it. In some places, huge baobab trees stand, their bloated trunks serving to store water soaked up during infrequent rains. Elsewhere low rocky hills stud the landscape. Many of the rivers are flanked by long galleries of forest, for water soaks through the soil on either side of their beds and allows trees to grow. But almost everywhere there is grass. In parts it grows higher than a man. Elsewhere it is low and so sparse that great areas of red dusty soil show between the clumps.

This varied landscape supports a varied population of animals. The chains of hunters and hunted are here just as they are in other grasslands, though the actual species, in nearly all instances, are unique to Africa. The grass is cropped by termites and ants. These are fed upon by specialised ant-eaters – pangolins and aardvark – as well as by more generalised insect-feeders – mongoose, and a multitude of birds. Small hunters – weasels, genets, jackals – prey upon vegetarian rodents – giant rats, spring hares and ground squirrels. Big carnivores – lions, hunting dogs, cheetahs and hyaenas – feed on the big herbivores. It is these large grass-feeders, nearly all of them ruminants, that dominate the African plain.

There are small ones like Thompson's gazelle and impala and big ones like eland, roan and topi. There are specialised ones, like giraffe that can nibble thorny branches well beyond the reach of any other browser, and sitatunga which live in marshes and reedbeds and only move out on to the plains when there are floods. And there are the non-ruminating giants, rhinoceros and elephant. Here herds still assemble in numbers that recall the tales brought back by travellers from the veldt and the prairie 150 years

The wildebeest migration, Kenya; Overleaf: *Crossing a river*

ago. Some species even still migrate in vast numbers to find better pasture when the seasons change, just as saiga and springbok and bison did in the past.

The most famous of these journeys is that of the wildebeest. Rain does not fall uniformly across the Serengeti: the southeast section dries out rather more quickly than the northwest. By May its grass has been cropped low, so its inhabitants have to move. A million wildebeests accompanied by zebras and gazelles begin a long trek, plodding in columns many miles long up towards the northwest. They plunge across rivers in stampedes and in such numbers and concentrations that many of them drown. More are forced into the water by the press of the multitude coming from behind. Lions ambush them and pick off exhausted travellers with ease. So they march day after day until, after 200 kilometres or so, they reach the still lush pastures of the Mara in southern Kenya. There they will stay and feed. But in November these plains too are beginning to fail and down in the Serengeti rain is beginning to fall once more. So once more the wildebeests must set out on their long journey.

Another less widely known migration is made by the white-eared kob, farther east in the Sudan. They move, not because of drought but because of flood. About a million of these handsome antelope, the males armed with graceful lyre-shaped horns, live on the grassy plains in the south of the country. Here, the does give birth to their young during the rainy season. As the rains end and the plains begin to dry out, they move north, following the new flush of green grass that springs from the receding waters. Their territory is flanked on either side by two rivers, swollen by the rains. Not far from the Ethiopian border, the two meet, and the kob, forced closer and closer together, are compelled eventually to try and make the crossing. And here the Murle people await them each year. In only a few days, 5000 kob may be killed. For the hunters it is an annual bonanza that provides them and their families with full stomachs for several months. For the kob it is the last ordeal before reaching the northern swampy grasslands and good pasture during the critical months of the dry season.

Ruminants are today the most successful of the large grass-eaters. They have far outstripped their only major competitors, the horses, in both the variety of their species and their absolute numbers, even now when their populations have been so heavily reduced by man. The shape of their bodies has been largely determined by the particular character of grass. The openness of the plains on which grass grows requires that those that live there shall be able to run swiftly to escape predators. Over many generations, the ancestral ruminants acquired that ability. They rose up on their toes and so developed longer legs. The lateral toes dwindled, the central ones strengthened and the nails at the tip thickened and provided durable shock-absorbing hooves. The seasonal sprouting of the grass, caused by the irregular rainfall which occurs on so many plains, necessitated long journeys to find pasture throughout the year and the animals, if they were to survive, had to be of a big enough stature to undertake them. Their stomachs changed into the large many-chambered kind that is able to digest grass efficiently, and their teeth too altered. Since grass grows close to the ground, those that eat it can hardly avoid taking into their mouths a certain amount of grit and sand. That, coupled with the intrinsic toughness of the grass leaves, causes heavy wear on the teeth and ruminants have developed very large grinding molars with open roots to them that can therefore grow throughout the animal's life.

But the influence has not all been one way. The ruminants have also affected the spread and extent of grass. If fire destroys a woodland in well-watered country, or man cuts down the trees, grass may establish itself. But seedlings also sprout, and within a year or so could create shade, the one condition that grass cannot tolerate. So woodland would soon displace grass and reclaim its own territory. With the help of ruminants, however, the grass's invasion can become a permanent occupation, for the animals will graze and trample the young tree seedlings and kill them. Only grass can survive such punishment.

But even grass requires some rain. As you continue north through the African savannahs, the rainfall diminishes and the land is drier. The thorn bush becomes more and more scattered and the grass thins. No more can you hope to see great herds of antelope. Even animal tracks in the dry sand at your feet are rare. You are now approaching another world, the desert.

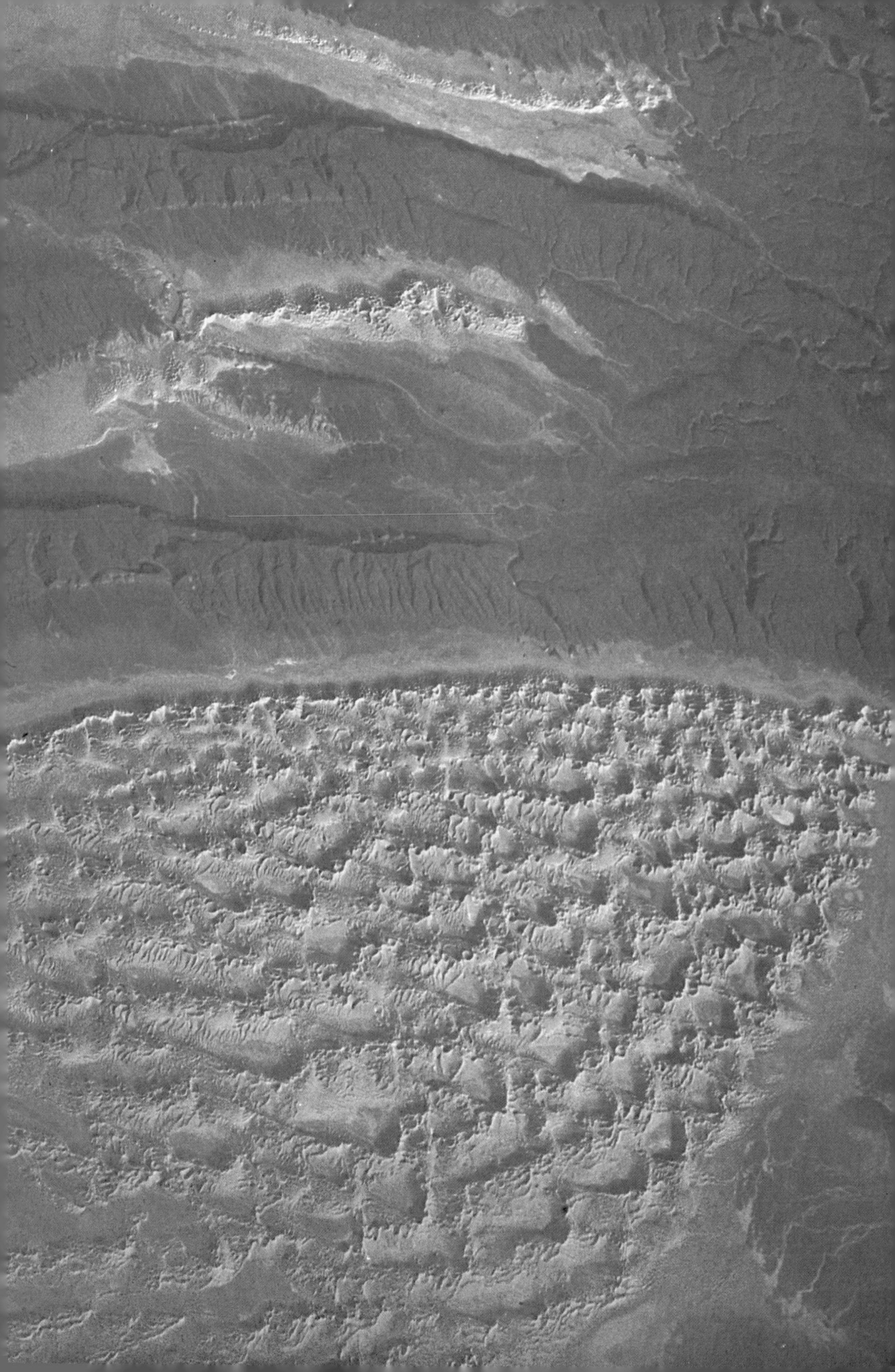

SIX

THE BAKING DESERTS

The Sahara is the greatest desert on earth. It stretches from the scrublands of the northern Sudan and Mali to the coast of the Mediterranean where its sands blow across ruined Roman cities. On the east it crosses the Nile and meets the waters of the Red Sea. Five thousand kilometres away to the west, it reaches the Atlantic Ocean. No river rises in it. Rain may not fall on parts of it for years on end. Here the highest shade temperatures on earth have been recorded: 58°C. Some of it is covered by sand. Much more is an arid plain of wind-polished gravel and tumbled boulders. And at its heart stand ranges of grotesque sandstone mountains.

They rise vertically from the top of a plateau, Tassili n'Ajjer, and form a bewildering tangle of precipices, rickety spires and curved archways. Most are more like tower blocks than mountains. Many, at their base, have been undercut to form shallow caves. Smaller columns have been sculpted into lopsided mushrooms. All these extraordinary shapes have been carved by the wind, whipping up the gravel and sand and blasting it at the surface of the rocks, cutting horizontal grooves in the cliff faces, deepening still further the lines of weakness between the layers of sandstone. The naked roasting rock, unprotected by vegetation or soil, is crumbling before your eyes, producing more sand which, in due course, new gales will hurl at the cliffs before finally carrying it away to leave it in piles elsewhere in the desert.

But the shapes of these mountains cannot all be attributed to the wind. The valleys between the towers follow courses like those of river valleys elsewhere and have smaller tributary ravines joining them, down which streams might have once tumbled. The implication that this land was once well-watered is strong, and further evidence is emblazoned on the rocks themselves. Beneath the overhangs, on the walls of the shelters, painted in glowing ochres of red and yellow are pictures of animals – gazelle, rhinoceros, hippopotamus, sable antelope and giraffe. Domestic animals are represented too – herds of piebald cattle with elegant curved horns, some with collars round their necks. The artists have also portrayed themselves, standing among their cattle, sitting beside huts, hunting with bows in their hands, dancing with masks on their heads.

We do not know exactly who these people were. They may have been the ancestors of

Satellite view of Sahara dunes

Rock painting of cattle and herders, Tassili

the nomadic people who still follow the herds of half-wild, long-horned piebald cattle that roam through the thorn scrub just beyond the southern fringe of the desert. Nor has it been determined precisely when the drawings were made. Several distinct styles can be identified and it is therefore likely that the drawings span a considerable period. Most experts believe that the earliest of them were drawn about 5000 years ago. But there is no doubt whatever that the scenes they represent can no longer be found in the surrounding desert. None of the animals, so vividly drawn, could live today in the hot bare sand and gravels of the Sahara.

Amazingly, one living organism has survived from that time. In a narrow rock-walled gorge stands a group of ancient cypress trees. Judging from the number of rings in their trunks, they are between 2000 and 3000 years old. They were saplings at the time that the last pictures were being painted on the rocks nearby. Their thick, twisted roots have pushed their way through the sun-riven rocks, widening cracks and tilting boulders as they have groped downwards for underground moisture. Their dusty needles manage somehow to be green and bring the only patch of that colour to the browns and rusty yellows of the surrounding rocks. Their branches still produce cones with viable seeds within them. But none germinate. The surrounding land is simply too dry.

The change in climate that turned the Tassili plateau and the whole of the Sahara into desert was very long drawn out. It started around a million years ago when the Great Ice Age that had gripped the world began to wane. The glaciers that had spread down from the Arctic, covering the North Sea with pack ice, grinding their way as far down as southern England and Germany, began to retreat. To begin with, this brought wetter conditions to this part of Africa and the Tassili became relatively verdant, but about five thousand years ago, the rains moved southwards and the Sahara became drier and drier. Its cover of grass and bush withered and died. Its shallow lakes evaporated. Its populations of animals and people wandered south in search of water and pasturage. Its soils blew away. Eventually what had once been a vast and fertile plain, studded with wide lakes, became a waste of bare rock and drifting sand.

It was not the first time that this had happened. Just as the ice-sheet over northern Europe had expanded and contracted several times, so the Saharan plains had oscillated between periods of fertility and aridity. Even so, this great segment of Africa, like all land in these latitudes, worldwide, south of the equator as well as north of it, has always been prone to drought.

The reason why rain does not fall uniformly over the surface of the earth derives, ultimately, from the unequal way in which the sun warms the earth, weakly at the poles and intensely at the equator. Hot air currents rise at the equator and then flow north and south into cooler latitudes, where they descend. Warm air can carry more moisture than cold air, so the rising equatorial air currents are initially very humid. But as they rise, they cool; their moisture condenses as clouds and finally falls as rain. The high altitude air, having shed its moisture, flows away towards the Tropics, 1500 kilometres north and south of the equator, and eventually begins to descend. By now it has lost all the water it once carried, so it brings no rain to the land beneath. Furthermore, as it

Overleaf: *Rock mountains of the Tassili plateau, central Sahara*

nears the earth's surface and is rewarmed, it sucks up any available moisture from the land over which it flows on its way back towards the equator. This circulation of air thus creates belts of parched lands around the Tropic of Cancer in the north and Capricorn in the south. These zones are not geometrically regular because the earth, spinning inside its atmospheric envelope, produces vast eddies in the air above and these are further distorted and complicated by the irregular distribution of land and sea, mountain and plain, on the surface of the globe. Even so, the broad pattern remains. Wherever land straddles the equator, there are pairs of deserts, north and south. The Sahara is matched, south of the rain-soaked forests of central Africa, by the Kalahari and the Namib. The Mojave and Sonoran deserts in the southwest of the United States have their equivalent in the Atacama in South America. And in Asia, the vast deserts of Turkestan and central India are paralleled on the other side of the jungle-covered lands and islands of southeast Asia by the great deserts of central Australia.

The absence of clouds above deserts has a double effect. Not only does it mean that no rain can fall, but it also denies the land beneath any shade from the sun during the day, and at night any blanket to help it retain its heat. While the desert during the day is as hot as anywhere on earth, at night it can drop below freezing point. Such great variations in conditions every twenty-four hours impose great problems on those creatures that have made their homes in deserts.

Most deal with them quite straightforwardly. They avoid, as far as possible, the temperature extremes. Small mammals shelter throughout the day, hiding in the darkness beneath boulders and down burrows. Inside such refuges, it is significantly cooler than in the glare of the sun, and the humidity, due in part to the occupants' own breath, is several times higher than it is above ground, so the animals lose very much less moisture. And here they stay for most of the day. Their time for activity comes when the sun sinks below the horizon.

In the Sahara, mouselike gerbils and hopping jerboas timidly venture out as soon as darkness falls. They are vegetarians. Tufts of grass are few, but they do exist and there may also be seeds and other bits and pieces of dead vegetation blown into their territories by the wind that will provide them with their minuscule meals. Geckos run in fits and starts across the cooling rocks, looking for beetles and other insects. Mammalian hunters appear too. Fennec foxes, their large triangular ears cocked and alert to every sound, flit silently over the rocks, nose to ground, checking the trails of scent that can reveal when who passed where. One may lead to a little gerbil. A quick pounce and the fennec has had its first meal of the evening and the gerbil its last. Caracals, a kind of cat, and striped hyenas also appear, as if from nowhere, and, in many parts of the deserts of the Middle East, there are wolves, smaller and with lighter, thinner coats than their more familiar relatives from farther north. In the deserts of the New World, similar gleanings and killings go on, though here it is kangaroo rats that hop after seeds and kit foxes and coyotes that hunt them.

After the initial pangs of hunger are appeased, there is a drop in activity. The temperature continues to fall. The geckos, as they lose their warmth, retire to their

Fennec fox, Sahara

crevices. The mammals, which generate their own heat internally, can continue their foraging and hunting even when the night has become quite cold, but even they have usually retired to their dens and burrows well before dawn.

When the sun reappears on the eastern horizon, a new set of animals materialises. This is the time, in the deserts of the American west, when the Gila monster starts its hunt. Apart from a closely-related species in Mexico, the beaded lizard, this is the only venomous lizard in the world. It is about a foot long, with a stumpy tail and a coat of shiny, bead-like scales, some coral pink, some black. In the early dawn, it moves only slowly. But as the sun warms its body, it becomes more and more active. Now it will snap up insects, birds' eggs and nestlings. It will even force its way boldly into a nest of desert mice and seize adults as well as young. In Australia, the little Moloch lizard, only a few inches long, comes out to feast on ants, sitting beside one of their trails, methodically snapping them up with flicks of its tongue, while the ants march doggedly and unheedingly onwards. Tortoises, too, in deserts everywhere, move out from their scrapes and holes where they have passed the night in the security of their shells.

Once again, this burst of activity does not last long. After a few hours, the sun has risen so high that the desert begins to bake once more. Reptiles can suffer from overheating just as mammals can, and four or five hours after sunrise, it is too hot even for them. Now the air shimmers over the stones. Rocks are painful for human beings to touch. The air is so dry and so hot that a man's sweat evaporates before he even notices it. In an hour he can lose a litre of liquid from his body. If he stayed out in the open for the whole day, without drinking, he would die. Even the slightest muscular movement generates heat. Now nothing moves unless it is forced to do so. And all the time, the sun beats down unrelentingly from the brazen sky.

The heat threatens plants just as it does animals. They also will die of thirst if they lose too much water through evaporation. The desert holly grows out in the open in the American deserts where there is no shade whatever. It reduces the amount of sunlight that falls on it by growing its holly-like leaves at 70° to the vertical, so that, for most of the day, the sun strikes only the edges of the leaves. Only in the mornings, when the sun is low and cool, do the rays shine directly on the surface of the leaves, supplying them with the energy they need for photosynthesis. The holly's leaves also excrete salt, absorbed from the ground and transported in its sap, and this forms a fine white powder on the surface of the leaf which reflects some of the heat in just the same way as the white clothes of an athlete do.

A few animals also remain above ground in the midday sun. The Kalahari ground squirrel uses its bushy tail as a parasol, erecting it over its head, hairs spread, and cocking it into just the position needed to keep its body in shadow. Others cool their blood by using radiators. The jack rabbit in America, a hedgehog in the Gobi and a bandicoot in Australia all use the same device as the fennec in the Sahara: huge ears. Clearly large ears are helpful in catching every sound in the desert, but the ears of all these creatures are big beyond the needs of acoustical duty. Through them runs a network of tiny blood vessels which are so close to the surface of the skin,

Jack rabbit, North America

back and front, that air blowing across them cools the blood that runs through them.

Other animals enhance the wind's cooling powers by using a liquid. The physical process that occurs when a liquid changes to a gas absorbs heat. So as water evaporates, it draws heat from its surroundings. This is why sweating cools a mammal. A similar effect is produced by panting. Air is drawn back and forth over the moist lining of the mouth so that saliva evaporates and blood in the tissues beneath is cooled. Tortoises, when they get really hot – above 40.5°C – wet their heads and neck with a great flow of saliva. Sometimes they go even further and release the large volume of liquid that they habitually store in their bladder all over their back legs. Kangaroos in Australia have developed a special network of capillaries close to the skin on the inside of their forearms. When the heat becomes intense, they energetically lick their saliva into a lather in the fur immediately above, and the process of evaporation draws heat from the blood beneath.

Birds are better equipped than most animals to keep out the heat. In most parts of the world, of course, their feathers serve to retain their body warmth. But insulators reduce the transmission of heat through them, no matter in which direction, and keep external heat out just as effectively as they keep body heat in. Protected by their feathers, many birds can sit unharmed in the desert sun throughout the day, but even they, on occasion, need to cool themselves, and do so with a more efficient form of panting than mammals use. They flutter their throats. This avoids the muscular effort of heaving the chest, but nonetheless creates a current of air across the moist inside of the mouth.

Sweating, panting, throat-fluttering and licking – not to mention dumping your entire reserve of urine – may all be effective methods of cooling, but a desert animal that uses them pays a high price. It loses that most valuable of all commodities, water. All desert organisms, animal and plant, go to great lengths to conserve the liquid in their bodies. Their droppings are usually extremely dry. A camel's dung can be used to fuel a fire almost as soon as it is produced, and that excreted by many reptiles is no more than a cake of dry powder. Even the use of water as a way of getting rid of soluble waste such as uric acid is employed only with great economy so that, whereas human urine contains 92 per cent water, that of a kangaroo rat contains only 70 per cent. One of the lizards in the Sahara even manages to get rid of its excess salt by excreting it through a gland in its nostrils.

The search for water dominates the lives of many desert creatures. A very few have so reduced their requirements that they can extract enough liquid from their food to survive and may never drink at all. Fennec and jackal obtain it from the body liquids of the animals they kill, the Dorcas gazelle from the sap of leaves, and kangaroo rats from seeds. One or two are able in great emergencies to produce water internally by breaking down their reserves of fat. But many big mammals, such as oryx and kangaroos, are doomed to trek daily from their grazing grounds to one of the few scattered water-holes and back every day.

Desert-dwelling birds often follow the same daily routine. Their problems become

Road runner feeding chicks, Arizona

acute during the breeding season, for their chicks need water just as much as adults and if the food they are given is not juicy enough, then liquid must be supplied in some other way. The sand grouse of Africa often nests as much as 40 kilometres from standing water. The cock bird ferries drinks for his chicks across this distance in a unique fashion. When he arrives at water he first drinks for himself. Then he paddles into the margin of the pool and stands very upright, deliberately soaking his belly feathers. These, on the male bird alone, have a structure found on the feathers of no other bird and they absorb water as a sponge does. Once they are fully saturated, he takes off for the nest. As he lands, his chicks cluster around him, necks craned upwards, and suck his feathers like puppies pulling at the teats of a nursing bitch.

The road runner, that jaunty, snake-hunting bird, often seen racing on its long legs across the deserts of Arizona and Mexico, has a different way of providing liquid for its chicks. A pair will build their nest in a cactus or a thorn bush and produce broods of two or three nestlings. These youngsters, from a surprisingly early age, are able to digest lizards and insects. When a parent arrives at the nest with a dead lizard in its beak, it does not hand over its catch straight away. A nestling begs, its beak agape, and the parent thrusts the lizard into its mouth. But it does not let the lizard go. While the pair remain locked, apparently in a static dispute over the prey, liquid appears from the back of the adult's throat and trickles down its beak into the youngster's mouth. This is not water that has been gulped down a few minutes earlier and stored temporarily in the adult's crop. Indeed, there may be no standing water within range. It has been produced in the parent's stomach by the physiological processes of digestion. Only after the nestling has, willy-nilly, taken its proper dose of water, is it allowed to eat meat.

The problem of collecting water from a near waterless environment must also be solved by desert plants. Few deal with it more efficiently than the creosote bush that grows in the deserts of the American southwest. It relies not on deep ground water, which in many deserts is beyond reach, but on the tiny film of moisture derived from dew or, exceptionally, left after a fall of rain, that surrounds rock particles several inches below the surface of the soil. It gathers this with a network of rootlets that penetrates the gravelly soil so far and so thoroughly that seemingly every available molecule of water is extracted. Each bush needs a large area of ground to supply it with sufficient moisture and once one is established in a really arid area, it gleans water so efficiently that no other plant can grow within several feet of it. This applies not only to other species of plant but to its own seedlings. So an individual bush tends to colonise nearby ground, not by setting seed and producing new individuals close by, but by sending out new stems around its base drawing on its slowly expanding network of roots. As the bush spreads outwards, so the stems in the middle tend to die away and the bush expands into the shape of a ring. With nothing else competing with it, the bush continues to grow outwards and the ring grows larger and larger. Some are now 25 metres across. The individual stems in such rings are not, in themselves, very old, but the plant, considered as a single organism, may have been growing and expanding

Ancient ring-shaped creosote bush, Mojave desert

on this one site for 10,000 to 12,000 years. And that makes the creosote bush the oldest known living organism in the world.

Some desert plants have adopted a different strategy for water collection. They do not, like the creosote bush, absorb tiny amounts more or less continuously but rely instead on the torrential downpours that come once every year or so and then absorb as much as possible as quickly as possible and store it. Cacti are specialists in this technique. There are some 2000 different species of them and all the ground-living ones live, in a natural state, only in the Americas. One of the biggest is the saguaro. It can grow to nearly 15 metres in height, forming either a single pillar or branching into several vertical fingers. Long grooves, like pleats, run up its length. When a storm does come, the saguaro sucks up rain water from the drenched ground, expanding its pleats as it does so, and considerably increasing its girth. Within a day, a large saguaro is able to take in a tonne of water. It then has to retain it.

The enemy now is evaporation. Water vapour is inevitably lost through the stomata on the leaves, so many plants in the parched and roasting desert, like those enduring the drought caused by freezing temperatures in the north, have very small leaves with relatively few stomata on them. The saguaro and other cactuses have gone one stage further. They have reduced their leaves to spines. The stomata have developed instead on the swollen stem which has become green and taken over the work of photosynthesis. The spines do more than protect the plant from grazing mammals, of which there are, in any case, very few. They break up any air currents that might blow around the plant so that the saguaro is, in effect, permanently surrounded by an invisible jacket of still air. The stomata are further shielded from draughts by being placed in the bottom of grooves just as they are on pine needles. And in addition to all this, the cacti have developed a special kind of chemical process that allows them to transpire, exchanging carbon dioxide for oxygen at night, when it is cool, and keep their stomata closed for most of the day. With all these devices, the saguaro is able to reduce its water-loss from evaporation to a minimum and retain most of its water for year after year, using it gradually to build new tissue, until there is another storm and another chance to refill its immense tanks.

A traveller in saguaro country, racked by thirst, might well be tempted to raid these huge water stores standing all around him. He would be very unwise to do so. The sap of the saguaro contains a powerful poison from which he might die. But this is not true of all water-storing plants. Indeed, both the Aborigines of central Australia and the Bushmen of the Kalahari rely on such plants for water during times of drought. Such desert-living people are expert botanists, putting to shame many an academically-trained biologist. I once followed an Aboriginal companion through the red desert of central Australia in search of water. He walked swiftly and confidently, not visibly searching by craning his neck from side to side as I was doing, nor even, apparently, casting about with his eyes. It was as though he were able to take in all his surroundings, the tiny blurred tracks in the sand, the shape of the rocks, the details of plant stems and leaves, with one penetrating glance. And then, unerringly, he knelt beside a short,

Saguaro cactus, Arizona

straggly stem with no more than a couple of small leaves hanging from it. To my eyes, it seemed identical to many other stems we had already passed. But to him it was clearly crucially different. With swift, powerful thrusts of a stick, he dug the soil away around it, following it down until the hole was a foot deep. There the pencil-thin stem suddenly expanded into a spherical root the size of a football. Grated pieces of it, squeezed in our hands, produced trickles of liquid. It was enough to slake our thirst. It could have been enough to save our lives.

The Bushmen of the Kalahari in southwest Africa have a similar expertise. Several different plants have such water-storing roots, but not all provide equally good drinks. The liquid from some is so exceedingly bitter that even the Bushmen are unable to swallow it. But it is not wasted. They use it for wiping their face and bodies, so bringing a welcome moistening and cooling of the skin.

The Bushmen are the only human beings who seem to have developed a specific anatomical adaptation to desert-living. All human beings store food reserves as fat. But a layer of fat that encircles the abdomen and parts of the limbs, as it does in most of us, is a great disadvantage in a desert. It makes it very hard for the body to lose heat through the skin, so the traveller, if he is generating heat in his muscles by moving, finds it very difficult indeed to stay cool. Some individual Bushmen, usually women, prevent this effect by having their reserves of fat concentrated in their buttocks, which become huge and contrast most markedly with the rest of their lean, stringy bodies. Their appearance may look odd to outsiders. In fact, it should be a source of envy to any fat sweat-drenched traveller of another race who ventures into the Bushman's desert.

The interlinked problems of keeping cool and retaining water face animals and plants in all parts of all deserts. But deserts are not uniform. Some areas have particular problems or particular resources that must be solved or exploited in various specialised ways.

The Namib, north of the Kalahari, has one source of moisture given to few other desert areas. It borders the coast. On many nights of the year, fog rolls in from the sea. As it flows over the desert, it condenses into droplets. Several of the organisms that live in the Namib depend upon it. On such evenings, darkling beetles, long-legged and black, clamber to the crest of the sand dunes and there stand in line, facing the coast head-down, their abdomens lifted high into the air, slowly marking time with their feet. The fog drifts past them, and drops of moisture condense on their bodies. As they lift their legs, it trickles down their thighs, on to their abdomens, and then at last down to their mouths where they sip it.

Fog also provides moisture for one of the Namib's unique plants, and certainly its most spectacular, Welwitschia. It has a large swollen root, rather like an immense turnip. In ancient specimens, this can be as much as a metre across at the top and project several feet above the ground. From its scarred and often contorted summit spring two, and only two, huge strap-like leaves. At their growing point on the root-top, they are green, smooth and broad, stretching from one side of the root-top to the

Darkling beetle collecting moisture, Namib desert

other. They curve upwards, like gigantic grooved shavings from the blade of a plane, and then fall in twisted, splitting coils to the ground. Their ends are frayed and withered where the wind has rubbed them back and forth over the stony earth. Were it not for that, the Welwitschia leaves would easily qualify as the longest in the world, for although they grow only slowly, the plant can live for over a thousand years. An unworn leaf from such an aged specimen would in theory be several hundred metres long. The great size of these leaves is at first sight an anomaly. Most desert plants, after all, have tiny leaves to minimise water-loss. But the Welwitschia's leaves, far from losing water, collect it. Just beneath their waxy surface lie groups of thin fibres that run the length of the leaf. They are markedly absorbent. When dew falls, molecules of water are first absorbed by the skin and then drawn even further into the leaf by the fibres. Other droplets trickle down the leaves and drop off their tattered tips to be collected by the plant's roots.

In some deserts, torrential storms occur with sufficient regularity each year or so to allow communities of animals to develop which, in effect, crowd their entire active lives into the brief period when water is relatively abundant. For most of the year, and sometimes for several years at a stretch, they are hidden and inert. A traveller in the desert can see no hint of the richness that lies all around him.

The first drops of rain are the triggers which bring it to life. Some may fall on clumps of withered dead plants, their leaves brown and fraying into dust, their stems topped by seed heads, dry and brittle. But suddenly they appear to come to vigorous life, furling back the brown covers of their seed-heads and exposing the seeds within. Others shoot their seeds several feet into the air. But the impression of life is quite illusory. The motive power for these actions is purely mechanical. As the rainwater is absorbed differentially by particular sections of the dead tissues, tensions are created which cause some parts to curl and others to release the seeds with a series of tiny explosions. But now the seeds themselves, lying on the ground, begin to move. As they also absorb water, the hairs with which they are covered begin to swell and stiffen, raising the seed upright into such a position that its first rootlet will strike straight downwards into the ground.

There is a possible danger in all this. It could be that these first drops of rain are a false start, an initial short shower, and the really heavy storm will not arrive for a week or so. If that were to be the case, then those seeds that germinate now will die in the dry days that follow. Some plants, however, have guarded against even this hazard. The coats of their seeds contain a chemical inhibitor that prevents germination. Only if the rain continues to fall sufficiently heavily, and long enough, to saturate the ground, will the inhibitor wash away and the seed be able to germinate.

As the rain soaks into the soil of the Arizona desert and the seeds begin to sprout, so the ground itself begins to stir. The surface of the earth cracks and small toads struggle out into the daylight. They are spadefoot toads and they have been lying buried a foot or so below the surface for the last ten months. The rain, swilling across the surface of the desert, has accumulated into shallow pools. The male spadefoots hop swiftly down

Welwitschia, Namib desert

towards them. Once in the water, they begin to call. Within a few hours, attracted by the urgent chorus, the females have joined them. They couple almost immediately.

Everything now proceeds with frenzied speed. Those toads that fail to meet their deadlines will not survive. Unless a toad finds a pool and mates on the first night of the emergence, it may not mate at all. Within a few hours the couples have laid and fertilised their eggs which lie in clumps in the tepid pools. The adult spadefoots have fulfilled their obligation to the next generation. Now they ignore their eggs and one another and begin to feed as fast as they can to prepare themselves for the long waterless months of starvation that will soon be upon them.

The eggs, meanwhile, are developing at an extraordinary speed. A day later and the pool is full of tadpoles. They are not the only creatures to be wriggling in the warm turbid water. Fairy shrimp, tiny crustaceans less than a centimetre long, are swimming in swarms. They have hatched from eggs that may have been blowing with the desert dust for fifty years or so, travelling maybe for hundreds of miles from where they were laid by their long dead parents. The dust has also produced microscopic spores and these now, in the water, have grown into thin filaments of algae.

The tadpoles feed feverishly. The algae alone will sustain them, but if there are fairy shrimps in the pool as well, some tadpoles will grow in a slightly different way from their brothers. They develop huge heads with much larger mouths than the alga-feeders and start to feed on the shrimps. Not only that, they also seize and eat their alga-feeding brothers. And all the time, the pool is evaporating and shrinking, so that its population has less and less room in which to swim and less and less water from which to extract oxygen. As the pool shallows, so it also warms and this brings further privation, for the warm water carries less oxygen.

With two kinds of tadpole swimming together in the puddle, the spadefoot toad is prepared for different eventualities. If there is another shower of rain, the water in the pool will rise again and the need for the speediest possible development will become less urgent. The new water, however, will have stirred up the pool and made it muddy and turbid. In these conditions, the carnivorous tadpoles do not fare well. They find it difficult to see their prey. No such problem faces the browsers. They continue eating the algae and grow steadily. Eventually, they turn into froglets and leave the pool, with luck, in some numbers.

But if no more rain falls, then it becomes essential for at least some of the tadpoles to complete their development at the greatest possible speed. In the shrinking puddle, the cannibalistic tadpoles consume their brothers and compete with one another for the deepest part where the water will remain longest. Soon those on the margins do not have enough water to cover their bodies and the sun bakes them to death. In the centre, there is little more than liquid mud. But in it the biggest and the most aggressive of the carnivorous tadpoles, if they are fortunate, will develop legs and hop away into the desert. Many will be snapped up by lizards or desert birds, but some, after a few weeks of feeding, will find crevices and cracks where they can take shelter against the oncoming heat. Their parents too will be starting to dig themselves burrows with the broad

Top: *Tadpoles of the spadefoot toad*
Bottom: *Spadefoot toad burrowing*

powerful hind feet that give them their name. Once they are underground, the outer coat of their skin will harden to form a watertight wrapping that seals them totally, except for two tiny holes at the nostrils so that they can breathe.

The pool has long since dried out. None of the adult shrimps survive but their eggs are blowing about in the dust. Many of the tadpoles never completed their development. At the end, they lay, cheek by jowl, in a solid mass which the sun congealed and shrivelled. But their bodies were not wasted. As they decomposed, their substance seeped into the sand at the bottom of the hollow that held the pool. This is where the next rains will form a puddle, once again, and here in the sand there will be organic fertiliser to hasten the growth of the next generation of algae.

The benefits of the storm have not yet all passed. The seeds that began sprouting with the first raindrops have grown swiftly into plants and they are now in flower. Acre after acre of the desert is ablaze with colour. Blue and yellow, pink and white, the flowers stand in vast transient meadows. So the deserts in western Australia, in the Namib and Namaqualand, in Arizona and in New Mexico, for a few short days are as colourful as any plot of wilderness in the world. Then with their moisture absorbed and their seeds set, the plants shrivel and die and surrender their place once more to the sand.

The conventional image of the desert, however, is neither gravel pavement nor wind-carved mountains, but endless dunes of sand. In fact, dunes occupy only a small proportion of the deserts of the world, but they do provide the most specialised of all desert environments. The sand from which they are formed is all that remains of the desert rocks after thousands of years of being grilled by the sun during the day and chilled to freezing point during the night. Under such conditions even the most durable granite begins to crack and flake. Slowly it disintegrates into its constituent minerals. Each grain, as it is hurled repeatedly against the cliffs by the wind, blown over flat rocky pavements and rubbed against other grains, becomes rounded and coated with a red polish of iron oxide. As the winds bluster and eddy across the desert, so they gather the grains sweeping them into great piles. These are the dunes. Some are as much as 200 metres high and a kilometre across. In those parts of the desert where the winds are constantly shifting, such mountains of sand may become star-shaped with half a dozen ridges leading up to a central summit, and remain roughly in the same place for centuries so that they become landmarks with their own names by which travellers across the desert find their way. Where the wind usually blows from the same direction, the dunes are far from stationary. They form ridges, like ripples on the sea floor, and slowly advance across the desert. The wind blows the sand up the gentle slope of the dune to its crest. Then, with nothing to bind it, it slips down the steep front face of the dune in a continuous series of tiny avalanches and the dune itself inches forward.

Dune sand presents great problems to any creature that attempts to live in or on it. Keeping a foothold on a surface which is extremely hot and continuously slips downwards is not easy. Several creatures have developed special feet to assist them. One gecko in the Namib has webbing between its toes, like a frog. Another has long hair-like fringes around its feet, which, in a similar way, spread the animal's weight, so that it can skim over the surface with the minimum disturbance of the sand, and therefore the least slippage. When this one stands still, it performs what appear to be keep-fit exercises, regularly and rhythmically lifting alternate back and front legs. By doing so, it manages to keep its feet cool and allow what wind there is to blow around its body.

The dune surface can become broilingly hot within a few hours of sunrise. Yet coolness is only an inch or so away. Thrust your hand beneath the surface of the sand and you will feel how surprisingly cold it is. Most of the dune animals are also well aware of the fact and will duck down beneath the surface to hide or to escape the worst of the heat.

Life within the sand may be cooler, but it has other problems. The grains are so smooth and dry that they do not cohere. So it is impossible to tunnel, in the way that can be done in soil. The sand simply collapses behind the tunneller. One way to move through it is with a swimming action and several of the lizards that regularly dive beneath the surface push themselves through the sand with their legs. But the best way of swimming in sand is not to use legs at all but simply to wriggle. Several lizards belonging to the skink family do this. Their legs, very much reduced in size, are sufficient to move them over the surface but are held close to the body when they are moving within the sand. One or two that spend nearly all their time below the surface have lost their legs altogether. The legless skink of the Namib is only a few inches long and looks like a tiny smooth-scaled eel. Scaly transparent covers have grown over its eyes, protecting them from wear by the sand grains, and its nose is pointed, assisting it in moving between them. It lives by hunting beetle grubs and other insects. The tremors in the sand caused by a moving insect are detected by the skink. It swims through the sand towards the movement and then pops up to seize its unsuspecting prey.

The legless skink is itself hunted by a sand-living mammal, the golden mole. This creature must be among the least known of all mammals, so rarely is it seen. Normally the only indication of its presence is a wandering line of footprints across the dunes, where it has emerged during the night, and a puddle-like depression where it has suddenly dived beneath the surface. It is such a skilful and energetic tunneller that it is almost impossible to dig one out unless you are lucky enough to start the chase when the animal is still close to the top.

It is about the same size as a European mole and has a general resemblance to it, but in fact the two animals are not closely related. Their apparent similarity is due to the fact that both, on different continents, took separately to digging and evolved similar adaptations to the underground life. The golden mole's fur varies in colour between species. Some are grey, some a beautiful golden yellow with a metallic sheen. It has no external ears, its eyes are covered with fur and without function; its naked nose is expanded into a broad sharp-edged leathery wedge with which it pushes its way through the sand; and although it has not totally lost its legs, its limb bones are buried within its flanks so that only the feet project. Sometimes it comes above ground to hunt insects, but its favoured

Overleaf: *The desert in flower, central Australia*

prey are the little legless skinks which it pursues underground, blindly burrowing at speed.

Few men live in the dune deserts. There is nothing here for them – no animals to be hunted, no plants to be cropped. But they do pass through it. The Tuareg, who originate from the northern side of the Sahara, regularly lead caravans of camels across it, carrying ingots of bronze, cakes of dates and bolts of cloth down to the ancient trading cities of Timbuktu and Mopti on the Niger River and bringing back huge rectangular cakes of salt. They protect themselves from the burning ultra-violet rays of the sun by swathing their bodies in flowing cloaks and wrapping their heads and faces with turbans.

But even the Tuareg could not make their journeys through the dune desert without the aid of an animal – the camel. The origins of this creature are still uncertain. Although there may be small groups of truly wild two-humped camels still surviving in remote parts of the central Asian desert, there are no wild dromedaries, the single-humped camel of the Sahara. Even so, it is unlikely that the truly wild creature differed greatly from the Tuareg's animal. They are marvellously suited to desert travel. Their feet have only two toes connected by skin so that as they walk their toes splay out and, with a web between them, do not sink into the sand. Their nostrils are equipped with muscles so that the animals can close them during sandstorms. Their bodies are covered with thick coarse wool on the upper surfaces, where insulation is needed from the sun, and more or less naked elsewhere, so that they can easily radiate excess heat. They have an astonishing ability to eat the thorniest of desert plants. Their food reserves are stored, as in most mammals, in the form of fat. This is not distributed around the body where it might prevent them cooling themselves, but concentrated in one place, in the humps on their backs. With this supply, they can survive for many days. At the end of periods of starvation their humps are little more than floppy, skinny sacks.

The camel's most celebrated characteristic, of course, is its capacity to walk across the desert for days on end without drinking. They do this by taking on board great volumes of water before they start on their journey and storing it in their stomach. They are also able to convert part of their fat reserves into liquid. In this way they can travel without drinking anything whatsoever for four times as long as a donkey and ten times as long as a man.

Yet even a camel cannot travel through the dune country of the Sahara without the help of men. Without the Tuareg to haul buckets up well shafts and tip them into troughs, the camels would find no water at all in the sand and the crossing would be beyond their endurance.

The oases, which are the essential staging posts on the long desert crossings, obtain their water from a layer of water-bearing rock deep below ground. The water is used by the villagers to irrigate gardens that are an astonishing demonstration of how fertile all the desert could become, if only it could be watered. Peaches and grain grow on carefully tended plots. Dragonflies hawk above the gurgling irrigation channels and birds sing in the date palms. Yet just beyond loom the sand dunes, ever threatening.

Tuareg, central Sahara

One great sandstorm, one season of the wind blowing constantly from a particular direction, and the oasis would be overwhelmed and extinguished. That, in microcosm, is the history of the Sahara over the past million years.

The evidence of the Tassili paintings show how recently it was that a variation in world climate turned fertility into desolation and created the Sahara. There is good evidence that most of the deserts that exist elsewhere in the world today were also formed at around this same time. Many of the animals and plants were exterminated by the new parched conditions. Some managed to cling to their ancestral territories by doing no more than altering their habits. Wolves and hyenas, gerbils and mice that had lived happily in grasslands and savannahs managed to maintain their place by restricting their activities to the hours of darkness when the desert becomes reasonably cool. Other creatures had to modify their anatomy in order to withstand the crushing hostility of the heat and drought. They altered the processes of their internal chemistry. They changed the proportion of their bodies. Some lost limbs; others developed slightly different ones.

The time scale of evolution is an immensely long one, measured in millions of years. Seen in that perspective, the animals and plants that live today in the deserts of the world have achieved their adaptations with quite extraordinary speed.

SEVEN

THE SKY ABOVE

Once a permanent trickle of water is established in a desert, living organisms appear in and around it as if from nowhere. A green skin of algae spreads over the sand grains in its bed. Small shrimps and crustaceans paddle through it. Mosses and flowering plants sprout along its margins. Mosquitoes play across its surface and dragonflies shoot and zigzag in pursuit of them. All these plants and animals have arrived without any help from man, or indeed, any effort on their own part. The only characteristic they needed to reach here, on a journey that may have spanned hundreds of miles and lasted for years, is infinitesimal weight. They have been carried by the wind.

Land-living organisms have been using this worldwide transport system for at least 400 million years. Long before any animal crawled out of the water, mosses had begun to colonise the land. Soon after they emerged, they began to use the wind to spread to new sites, just as their direct descendants do today.

Living mosses produce their spores in small capsules at the top of their stems. As each ripens and dries, a lid springs off the top to reveal a ring of teeth covering the mouth beneath. If the weather remains warm, these teeth also dry out and begin to curl back, so opening the capsule and allowing the spores to be blown out by the wind. If the weather should turn wet, the spores would soon become waterlogged and could not be carried far, but they are not released in such conditions, for as the air becomes humid, the little teeth reabsorb moisture, straighten out and shut the capsule.

The number of spores produced by mosses is vast, but it is small in comparison with the truly astronomic quantities discharged by fungi. An ordinary field mushroom, when it is ripe, releases from its plate-like gills 100 million spores in an hour, and before it decays may have produced 16,000 million of them. A giant puff-ball exceeds even that number. An average-sized specimen about 30 centimetres across, according to one botanist's estimate, may produce seven million million. It shoots them into the air, a thousand million or so at a time, like puffs of brown smoke, every time it is knocked or given a slight buffet by the wind.

Simple plants are not the only ones to exploit the wind in this way. Those highly sophisticated and very complex organisms, the orchids, also do so. A single bloom may produce as many as three million seeds. Such dust-sized grains cannot carry food reserves as well as an embryonic plant, so an orchid seed, if it is to develop successfully, must land on a fungus similar in character to those that surround the roots of some trees, which will assist it nutritionally through the first stages of its growth.

Most higher plants, however, do provide a ration of nutriment with each of their seeds. This so adds to the weight of the seed that the wind is unlikely to be able to lift it into the air and carry it any distance without some device to increase its surface area. Thistles, bulrushes and willows equip their seeds with tiny tufts of down. The dandelion develops a filamentous parachute for each of its seeds which can easily carry one for 10 kilometres from the parent plant and often does so for many times that distance.

So all over the globe, the air contains tiny organic particles, many so small as to be invisible, bearing the germs of life. Most will never develop. They are collected and eaten by insects; they fall on sterile ground and rot; they are blown about for so long that their life is extinguished and they disintegrate. But one or two out of several million will survive and wherever a suitable vacant space appears – on a dead leaf or an untilled garden plot, in a trickle over a rock face or a desert puddle – a green plant or a fungus will appear. So mosses germinate in the oases of the Sahara and on isolated volcanic islands in the Antarctic seas, kapok seedlings sprout throughout the South American jungles, and willow-herb blooms on the bare ash fields of Mount St Helen's.

One or two animals are small enough to distribute themselves by the same method. The tiny brine shrimps of desert pools have hatched from dust-sized eggs that blow in the wind. Mosquitoes, aphids and other small flying insects are carried by the breeze for many miles whether they like it or not. Many young spiders however take to the air intentionally. When they emerge from their cocoons, they climb up a grass stem or the pinnacle of a pebble, face into the breeze and raise their abdomens. From the spinnerets at the end of their bodies, they produce a thread of silk. Even the slightest breeze will catch it and draw it out. As it extends, the wind gets an increasing hold on it. The infant spider resists the drag for a few moments, clinging fast with its legs, but then it lets go and sails off into the air. Such spiderlings, hanging from their threads, have alighted on ships in the middle of the ocean, hundreds of kilometres from land, and on snow peaks thousands of metres high. When the wind eventually deposits them on the ground, they detach their threads and start to establish themselves in new territory. During some times of the year, when the weather is suitable, huge numbers of such spiders may be deposited together, by a quirk of the breeze, in one small area. Their abandoned threads then become entangled and mat together to produce the once-mysterious fabric called gossamer.

Other small creatures, scarcely any bigger, also travel by air, but do so under their own power. Thrips are minute, sap-sucking insects which live on flowers, leaves and buds. In order to move from one plant to another, a thrips flies, but it is so small, so light and has such microscopic muscles that it is very difficult indeed for it to beat its wings. It is as though the air around it were viscous, like treacle. So the wings that sprout from its thorax are not broad blades but merely thin rods fringed with hairs. A downward

Puffball discharging spores

stroke from one of these increases the pressure in the air below it and slightly reduces pressure in the air above. In consequence, the thrips is sucked upwards and it takes off, like a fleck of motorised thistledown.

The creation of high pressure below a wing and low pressure above it produces lift. It is one of the basic forces on which powered flight depends. A bumble bee, many times heavier and stronger than a thrips, needs broad wings to produce sufficient lift. Beating such large structures requires considerable strength and the bee's thorax is packed with muscles. They, like other engines, have to be warm if they are to work at full power and provide the energy necessary to lift the bumble bee's body into the air. But the bee, like all insects, does not maintain a steady warmth in its body as mammals and birds do. It normally draws its heat from the sun. Yet even on mornings when the temperature is within a few degrees of freezing point, a bumble bee manages to fly. It does so by shivering its wings rapidly from side to side before it takes off, so generating heat within the muscles. It can even put its wings out of gear and run its internal engine until the temperature of its muscles are as high as that of a man's blood. Warmth being so valuable to it, the bumble bee, like many other large insects, has a coat of hair around its body which cuts down heat-loss. Dragonflies insulate themselves for the same reason, but do so with a series of air sacs inside the walls of their thoraxes. With these powerful motors at their disposal, insects have become consummate aeronauts. A honey bee can beat its wings 15,000 times a minute and a dragonfly reach speeds of over 30 kilometres an hour.

Two other major groups of animals have joined the insects in the air. Some 140 million years ago, birds evolved from reptilian ancestors; and, much later, about 60 million years ago, some insect-eating mammals gave rise to the bats. Both bird and bat have developed wings by modifying the forelimbs. The bat has an elastic membrane of skin stretched between four greatly elongated fingers, with the thumb free to serve as a comb and a hook with which to clamber around in the roost. The bird has retained only one of its fingers which has become long, strong and fringed with quilled feathers. It, too, has kept a vestige of its ancestral thumb as a small projection on the fore-edge of its wings which carries its own tuft of feathers.

Bats, since they roost by hanging upside down from their feet, have no difficulty in launching themselves into the air. They can simply let go and fall into it. Some larger fruit-eating species flap their wings once or twice to raise their hanging bodies into a flying position, but this requires little effort. Most birds, however, are walkers as well as flyers and the problem of overcoming the pull of gravity and hauling themselves up into the air from a standing start is much greater. The engine that gives them the power to do so is the massive bundle of muscles that stretches from the wing-joint down to the deep keel on the breast bone. The fuel it uses, oxygen in the blood, is supplied in great quantity by a huge heart. How exceptionally large this is can be gauged by the fact that the heart of a sparrow is twice the size of that of a mouse. The bird's body is wrapped in that finest of all natural insulators, feathers, and kept at a temperature that is several degrees higher than a man's, so its flying engine can spring into immediate and powerful

Bumble bee in flight

activity at a moment's notice. With this driving their wings, and with an upward spring from their legs, most birds are able to launch themselves up into the air with ease.

But the heavier a bird is, the bigger the wings it requires to sustain it in the air, and the greater the muscular task of beating them with sufficient force and speed to produce take-off. There is, however, another way of producing lift. If a wing has the right curvature on its upper surface, a current of air blowing across it will produce the necessary low pressure above and high pressure below. That current can be created either by the wind blowing across the wing, or by moving that wing swiftly through the air. It is best done in both ways simultaneously – by running into the wind.

The wandering albatross has the biggest wings of any bird – 3.45 metres across – and flapping them rapidly is virtually impossible. So to take off, it has to use to the full this second way of producing lift. Often, it nests on steep cliffs so that it can simply tumble into the air. Other species of albatross nest in dense colonies on low oceanic islands, but no matter how cramped they are and how great the demand for nest sites, they leave a clear strip of ground alongside and sometimes even right through the middle of their colonies. This is their airstrip and it is accurately angled along the direction of the prevailing wind. The birds queue up at the end of the strip, facing into the wind, like jet planes at a busy airport, and when their turn comes they run as fast as they can, slapping the ground with their large webbed feet, leaning forward and beating their immense wings as rapidly as possible. Eventually, as a result of their exertions and of the gale blowing across the surface of their outstretched wings, they get the lift they need, rise into the air and are immediately transformed into creatures of the greatest elegance and grace, soaring away over the sea. But were the wind to drop to nothing, they would have considerable difficulty in even getting off the ground.

Once airborne, the albatross exploits the wind to travel with a minimum expenditure of energy. Close to the surface of the ocean, the air currents are slowed by friction with the waves. The albatross remains just above this slower layer, about 20 metres above the water, travelling with the fast wind. As the bird slowly loses height, it glides into the lower layer and turns into the wind, using the momentum of its speed to give it lift and so carry it upwards into the fast airstream and restore its height. The extremely long, narrow wings that were so difficult to flap during take-off now prove their value, for the albatross can maintain this swooping soaring flight for hours on end without a single wing-beat. Several species of them live in the stormy near-frozen seas that surround Antarctica, where the winds blow continuously in an easterly direction. The albatrosses travel with them, soaring round and round the globe, swooping down to the water only to catch fish or squid. Year after year, they remain aloft until, at last, when they are seven years old, they are mature. Then they land on one of the small islands that lie in their path. For a few weeks they spend much of their time on earth. They dance to one another with outstretched wings, clapping their bills. They mate and they rear their single chick. Then once more they resume their effortless earth-girdling flights.

Those other accomplished gliders, the vultures of Africa, have no such steady and reliable winds to assist their flight. They exploit a different kind of air current. The

Long-eared bat

Overleaf: *Courtship of wandering albatross, South Georgia*

surface of the earth does not react to the heat of the sun in a uniform way. Stretches of grass and expanses of water absorb heat, so that air above them remains relatively cool. A patch of bare rock or earth, however, reflects heat and so creates above it a rising column of hot air known as a thermal. Every morning, vultures, perched in the low thorn trees where they spent the night, wait for the sun to rise and heat the land. As soon as a thermal begins to form, the birds laboriously make their way across to its base, flapping and gliding, not attempting to gain any altitude, until at last they reach the column of rising air. It catches beneath their wings and lifts them. Their wings, though large, are not long and narrow like an albatross's, but short and broad. This shape enables them to make tight turns and they spiral upwards, keeping all the time within the narrow column of rising warm air.

When they reach the top of the thermal, hundreds of metres above the savannahs, the vultures wheel effortlessly round and round, scanning the plains beneath for signs of a kill or a stricken, sickly animal. Tiring of one thermal, they may leave it and glide gently downwards for 10 kilometres or so to pick up another and then spiral upwards once more to a new observation post. In such a way, they may travel 100 kilometres a day over the savannahs searching for food. Once they see it, they glide down steeply and swiftly and come in to land with their wings tilted and their tails lowered to act as an air brake. Squabbling and fighting with one another, they gorge themselves on the corpse until their stomachs are so loaded with meat that they have great difficulty in getting into the air again. Usually they labour across to the nearest thorn tree and there perch for some time, digesting their meal before attempting to make their way back to a thermal and climb once more into the sky.

Few birds can rely on the upward lift of air currents to take them where they want to go, in the way that albatrosses and vultures do. Most travel through the air by making a rowing action with the outside half of their beating wings. Their tails, constructed from a fan of feathers that can be broadened and narrowed, raised and lowered, enables them to control their direction through the air. This apparatus is so efficient that birds have become the biggest of all living flying animals. The Andean condor weighs up to 11 kilograms, as much as a small dog.

Moving through the air at speed requires extremely sensitive navigational equipment in order to avoid obstacles, to catch prey in mid-air and, above all, to assess distance with the accuracy necessary for a safe landing. Nearly all birds fly, for the most part, during the day and they rely almost entirely on their sight. Indeed, they have the most efficient and sensitive eyes possessed by any animal. A hawk's eyes are actually bigger than those of a man, in spite of the disparity of their body size, and are eight times better than his at distinguishing detail at a distance. Owls, since they hunt at night, have sacrificed perception of detail in favour of sensitivity. Their eyes are gigantic, much larger even than they appear, for only the central cornea is exposed and much of the rest of the eye is covered by skin. They occupy so much of the front of an owl's skull that they leave almost no room for muscles. So the eyes are, in effect, fixed in their sockets. If an owl wants to look to one side, it has to turn its head and it has a neck of

White ibis, North America

unrivalled mobility to enable it to do so. The gigantic cornea and the huge lens that lies behind it gather so much light that an owl can see clearly in only one-tenth of the light that a man would require.

But even owls, if they are to see, need some light. In its total absence, no eye, however optically efficient, can operate. Yet two birds have techniques for finding their way about even under these conditions. Both live in caves. The oil-bird, or guachero, is a relation of the nightjar. Its most famous colony is in the great cave of Caripe in Venezuela. A few hundred metres from the entrance, the cave bends, cutting out all direct light from the outside. A little further on and it is pitch black, so that you have to use a torch. By its light, you can see the oil-birds sitting on ledges, at all levels on the cave walls, in between the curtains and pillars of stalactites. They are as big as pigeons and their eyes glow in the beam of your torch as they move their heads from side to side looking down at you in curiosity. The nests on which they sit are mere mounds of regurgitated food and droppings. On the rock floor, at the base of the wall, fruit seeds, voided in their droppings, have sprouted to form high thickets of pallid, spindly shoots.

The light of your torch will alarm them and many take off, swooping around you, shrieking and squalling, so that the whole cavern rings with their cries. But if you turn off your torch and wait quietly in the blackness, the birds will settle down and their alarm calls cease. But they are still flying and, above the soft swish of their wing-beats, you can hear a barrage of tut-tutting clicks. These are the signals by which they navigate. They deduce from the echoes the position of the rock walls and hanging stalactites, and even of the other birds flying in the air around them. The frequency with which they make these clicks increases as they approach an obstacle and its precise position becomes more important to the bird. They are able to detect the presence of objects about the same size as themselves by this technique, but nothing much smaller. This, however, is all that they need to do, for once they have made their way safely out of the cave, there is sufficient light in the night-time forest for them, using their large sensitive eyes, to find the fruit on which they feed.

The other bird to employ this echo-location technique is the swiftlet that lives in caves in southeast Asia. It is quite unrelated to the oil-bird, but it too produces streams of rattling clicks. They are much higher-pitched than the oil-bird's and they enable the swiftlet to detect smaller objects.

Complex and sophisticated though these two birds' technique may seem, it is crude in comparison with the refined version developed by those habitual nocturnal flyers, the bats. The sounds they produce are very high-pitched indeed, and far beyond the range of human ears. Although some people, particularly when young, can hear the faint squeaks of bats hunting on a summer's evening, the majority of signals that the bats use for navigation are even higher still. Bats produce them in streams at extreme speed, as many as 200 a second. This enables a bat not only to find its way, but to pinpoint a flying insect with accuracy.

The mastery of the air brings great benefits to those that have achieved it. Bats can fly without difficulty great distances every night to tap particular and temporary sources

Oil-birds

of food. They can snatch insects from mid-air, hover in front of blossoms to sip nectar and even grab fish from the surface of rivers. Even so, they are not as superbly competent and versatile as the birds. A lammergeier, a kind of vulture, after stripping a carcass, picks up the bigger bones, carries them to a considerable height and then drops them on a rock, so cracking them to expose the nutritious marrow on which it feeds. Small hunting birds, like kestrels and sparrowhawks, are able to hover on quivering outstretched wings, exactly matching their forward speed to that of the wind so that they hang motionless in the sky while they examine minutely the ground beneath them for the slightest movement that might betray the presence of a mouse or a lizard. The peregrine, the swiftest of all bird hunters, patrols high in the sky. When it selects a small bird beneath as its prey, it dives on it, with wings swept back into the position that offers least resistance to the air, reaching speeds of up to 130 kilometres an hour. It hits its victim while still in the air with a powerful blow at the back of the neck which kills it instantly. Its speed and force is so great that were it to strike a target on the ground in such a fashion, both hunter and prey would be destroyed.

Some birds indulge in aerobatics, seemingly for the sheer pleasure of doing so. Ravens will tumble and roll in the wind, apparently in play. Other birds show off their aerial skills as part of their courtship display. Greenshank rise 600 metres in the sky and there begin diving, twisting and turning, all the time singing loudly. Some birds – such as lapwing and snipe – have special feathers which vibrate as they dive to produce sounds as part of their courtship. Bald eagles and black kites, displaying to their mates, throw themselves around in the air, one bird swinging itself over on its back so that the pair, like circus acrobats, clasp claws in mid-air.

The greatest boon conferred by flight, however, must be the ability to make long journeys over land and sea, untroubled by all the obstacles that impede earth-bound animals. Birds fly from one continent to another, to avoid severe winters or to gather seasonal banquets of fruits or insects. Exactly how they do so, we still are not sure. It may involve finding their way by the sun and the stars, recognising the pattern of the land beneath them, and responding in some fashion to the electro-magnetic field of the earth. Less well-known are the migrations made by bats.

Many bats, in autumn, when the summer swarms of insects begin to disappear and cold weather threatens to chill their small bodies, seek caves in which to hibernate. Their requirements for a winter roost are exacting. It must be dry, not too cold and with a steady temperature. The number of such sites is not large and many species of bat that have ranged widely as they fed through the summer fly hundreds of miles in the autumn to a chosen cavern or loft. Others congregate for different reasons. In Bracken Cave in Texas 20 million free-tailed bats assemble every summer. All are female. They have left their mates 1500 kilometres to the south in Mexico to come here to give birth to their young. It may be that, since the young bats are born naked, the warmth generated in the cave by this vast congregation is of value to them, but we still do not fully understand the compulsion that leads all these female bats to gather in such immense maternity wards.

Lammergeier with bones

Insects also make long journeys by air, but because their flight seldom appears purposeful, naturalists were slow to recognise them for what they are. Butterflies that in summer flutter around their foodplants in meadows and woodlands seem to be so fragile and feeble that it is easy to assume that they will not stray far. Some species, indeed, do not. They feed, mate, lay their eggs and die in the same small patch of countryside where they hatched. Many other species, however, spend their lives travelling. A cabbage white butterfly, for example, hatched in spring somewhere in Europe will fly, for the most part, in a roughly northwesterly direction. It only travels when the sun is out and the day is warm. It will not go fast and it dallies on the way if it finds a patch of vegetation that suits it. There it may stay for several hours, feeding, courting or laying its eggs, but eventually it will move on. Its life is short – only three or four weeks – but even so, within this brief time it may travel 300 kilometres from the place where it hatched.

More cabbage whites hatch towards the end of summer. They, too, are travellers, but they move in the opposite direction, to the southeast. And those that appear in the middle of the summer begin by travelling northwest for a week or so and then, within a few days, change direction and spend the rest of their lives travelling southeast. The precise date of this turnabout differs according to the area and the species of butterfly concerned, but in one place, for one species, it is precise and constant from year to year. The factor that triggers the change seems to be the length and the temperature of the night.

Butterflies like these navigate by the sun and apparently make little or no allowance for its daily movement. This results in a very broad migration path, but since the function of their journey is not to reach a particular point, but to discover new feeding grounds, mates and egg-laying sites, such a track suits them very well.

A few species of butterfly make very different kinds of migrations. The most famous is the monarch. A large population of them inhabits the woodlands around the Great Lakes of North America. They are long-lived butterflies, some surviving for almost a year. Those that hatch in the spring are likely to stay in the same neighbourhood all their lives. In early autumn, another generation emerges. Some of these, too, will not stray far. Having fed, they will seek shelter in hollow trees or in the narrow space between a dead trunk and its bark, and there they hibernate. Two-thirds of this autumn generation, however, behave in a quite different way. They set off southwards. They follow a well-established and quite narrow route, flying in a purposeful way, seldom stopping to feed or court. Each night they roost, often occupying trees that have been used by many generations of monarchs before them. They, too, steer by the sun but apparently they know how to compensate for its daily movement, for their path is straight and quite unlike the meandering flight of foragers such as the cabbage white. Eventually, after a journey of some 3000 kilometres, they reach southern Texas and northern Mexico. Those that arrive in Mexico assemble in one or two specific valleys and go to roost in millions on particular coniferous trees that have been used for generations. So thickly do they settle that they swathe the trunks in a continuous fur of

Courting white-bellied sea eagles

wings. Others settle on the branches, hanging on to every available needle, so that they seem to be dripping with butterflies.

On warm days, a few of the millions may flutter a little distance and feed in a desultory way; but most of the time they all rest. Only when spring comes do they begin to stir. Until now, they have been sexually inactive, although in all other respects adult. But now they mate. Then, over a period of several days, this immense blizzard of butterflies begins to move northwards. This time they do not travel so fast. Usually they cover little more than 15 kilometres in a day. As they go, they feed and lay eggs. Few, if any, of these migrants will get back to the woods where they were hatched in the north, but they have left their offspring along the track of their journey, and next autumn more monarchs, emerging from eggs laid by residents of the north, will make the long journey south.

The height at which migrating insects travel varies. On windy days, butterflies keep low, taking shelter behind lines of trees, hedges and walls, in order not to be blown off their course. On fine still days, however, they may rise to 1500 metres above the earth's surface. Involuntary aeronauts, like the young spiders, are swept up by air currents to even greater altitudes. As they go, all these tiny travellers are preyed upon by high-flying insect-feeding birds such as swifts. Those that escape may rise as high as 5000 metres.

The character of this world only a few miles above our heads can hardly be appreciated from the pressurised, heated and oxygen-enriched cabin of a passenger aircraft. Float up to it instead in an open basket suspended beneath a balloon. For the first few hundred metres, noises from below, of car engines, snatches of chatter, a striking clock, sound distantly in a curiously unreal way. But soon all is silence, broken only by the creak of the basket and the sporadic roar of the burner that produces a blast of hot air and gives the balloon its lift. The atmosphere gets steadily colder. You are travelling with the wind, like many another creature swept up here by rising hot air, and therefore all seems still, even though you may be moving very swiftly indeed in relation to the earth beneath. But that may already be hidden from you beneath a bank of cloud. The air you breathe is rapidly becoming thinner and therefore each breath you take contains less oxygen. Since you are standing still in the cramped basket, this is not likely to trouble you. Indeed, you may be largely unaware that there is any change in the physical character of the air. This is what makes it so dangerous, for as your brain receives less oxygen, so it becomes less efficient and your faculties begin to dull. Long before you are aware of any physical impairment, you may have lost the ability to make competent judgements. So by the time your altimeter shows that you have reached a height of 5000 metres, you will be wise to start breathing oxygen through a mask.

The world you have entered is one of extraordinary beauty. Far below, a gauzy sheet of cloud may veil the surface of the land. Hills project through it, like islands in a white sea. All around sail great clouds, their lower margins flat and horizontal, but their upper surfaces billowing and surging in rapidly-changing plumes. As you approach their level, the extraordinary speed of the currents within them becomes vividly and

Monarch butterflies hibernating, Mexico

frighteningly apparent. To be caught by one of the thermals that feed them and be carried up into them would almost certainly be lethal. Within them, air currents are sweeping up and down with such force that they would rip apart the balloon. Above these clouds, there may be wisps of others at very high altitudes indeed and, beyond them, the clear dark blue of space.

It is just possible, at this height, that you might see other living creatures. Flocks of chaffinches have been recorded flying at 1500 metres; and shore birds detected by radar at an almost unbelievable height of 6000 metres. They are migrants and may have come to these heights to take advantage of the winds that are usually stronger and steadier than at lower altitudes, or – in those cases where the flight was at night – to catch sight of the stars by which they navigate. But these visitations are sporadic and rare. A few other creatures can be found here at most times. A protracted search, using slides covered with a film of grease, will produce a few aphids, one or two of the tiny gossamer spiders, and those ubiquitous specks of dormant life, pollen grains and fungal spores.

But this is the last frontier of life. No living organism of all the many millions that proliferate in the thin layer below comes any higher – except man. Another kilometre or so above this level and almost all the gases of the atmosphere come to an end. Beyond lies the black emptiness of space.

The swathe of gases through which we have climbed, insubstantial though it is, provides an invaluable shield against the lethal bombardments of space. Cosmic rays, X-rays, and the damaging elements in the rays of the sun are all absorbed by this gaseous blanket. Meteorites – fragments of stone and metal shot from outer space – are burned to dust by friction with the gases. On earth we see their arrival and extinction as the tracks of shooting stars. Only very few are big enough to survive and hit the earth. The atmosphere also protects us from extreme variations of temperature. How crippling they would be can be judged from the conditions on the moon which has no wrapping of gas. When the sun shines on it, its surface becomes so hot that water would simmer. In shadow, it becomes far colder than the lowest temperatures recorded on Antarctica. Here on earth the atmosphere absorbs much of the energy of the sun's rays as they pass through it, so our days are tolerable; and at night it prevents the warmth the earth has absorbed from escaping back into space.

By far the most abundant element in the atmosphere – nearly 80 per cent by volume – is a totally inert gas, nitrogen. This was probably discharged during the vast volcanic eruptions that racked the earth's surface during the early stages of its formation. The nitrogen has been held around the planet ever since by gravitational pull. Oxygen, which constitutes a little more than 20 per cent of the atmosphere, is a more recent addition. It has risen from the plants on the earth's surface as a by-product of their photo-synthesis. The remaining part of the atmosphere, less than 1 per cent, consists of carbon dioxide and some minute traces of rare gases such as argon and neon.

In addition to all these gases, the atmosphere also contains water. Some is invisible vapour, some the fine droplets that accumulate as clouds. Abundant though clouds appear to be in the sky, atmospheric water is only a minute fraction of that which lies

Clouds above volcanoes, Java

Hurricane seen from a satellite

below on the surface of the earth, in its ice-caps and snows, lakes and oceans. And it is from the earth that the atmosphere's water has come. Some has been given off from the leaves of plants but most has evaporated from the surface of the seas and the lakes. Sometimes the process takes place gently and evenly over great areas, producing layers of horizontal clouds. Sometimes the vapour is swept up in the great surges of hot air that rise as thermals to condense in the towering cumulus.

These great accumulations of water particles are blown over the surface of the planet by winds that are created by those two familiar factors, the earth's spin and its unequal heating by the sun. The first imparts an east-west surge to the air, the second a north-south movement as hot air rises at the equator and descends at the poles. As these two influences interact, they produce vast eddies. The clouds that develop above the warm waters of the oceans are swept into vortices that may measure 400 kilometres across and be so thick that they occupy all the atmosphere from top to bottom. The wind sweeping round these systems may reach speeds of 300 kilometres an hour. These huge storms are hurricanes, the most gigantic and catastrophic of all atmospheric disturbances. Torrential rains, driven by the fastest winds to sweep the earth, lash the land and the sea. The ocean is driven by gales into walls of water that sweep over the coasts. The screaming winds smash trees and rip buildings apart, while torrential rain is thrown down from the black racing clouds.

But for the most part, water in the sky falls in a gentler fashion. The cumulus clouds sometimes rise so high that their droplets turn to ice. In particularly large ones, that may measure 4 kilometres from the base to their crest, the ascending air may catch these fragments and blow them to the top of the cloud. As they go, they collect an additional glazing of ice, increase in weight, and fall, only to be caught again by an up-current. They may rise and fall many times before at last they are so big that they drop through the base of the cloud and fall to earth as hail. Less powerful clouds shed their ice particles while they are still very small. They melt as they fall and turn to rain. Layered clouds are swept over banks of colder, denser air, chill as they ascend, and so shed their load of moisture. Other clouds, blown across rising ground and on to the flanks of mountain ranges, again produce rain. So the fresh water, on which all land-living animals and plants depend, returns to the planet from which it came.

EIGHT

SWEET FRESH WATER

The snow flakes, falling so gently on the mountains of the world, are agents of destruction. They mantle the peaks in fields metres deep. Their lower layers, compressed by the weight above, turn to ice. It closes around projections of rock and penetrates the cracks and joints. As the snow continues to fall above, the ice beneath begins to move slowly down the steep slopes under its own weight, dragging away plates and blocks of rock as it goes. Most of the time, the movement is so slow that its only visible signs are widening cracks across the snowfield. On occasion, the whole sheet suddenly loses its grip and thousands of tons of ice, snow and rock sweep down the mountain.

All this frozen water, gathering in the vast couloirs between the mountain ridges, unites to form a river of ice, a glacier. Now the destruction becomes devastating. As the glacier slides downwards, it scrapes away at the sides of the valley against which it presses. Beneath, boulders frozen into its underside, like teeth in a gigantic rasp, grind down its bed. Ahead, it pushes a huge wall of shattered rock. Slowly it inches downwards below the level of permanent snow until the warmth begins to melt it, and water, creamy with pulverised stone, gushes from its snout.

Rain, falling on the mountain at these lower altitudes, is also a destroyer. During the day, it trickles harmlessly over the bare rock faces and permeates their crevices, but at night, when it freezes, it expands and wrenches away splinters and flakes which tumble down to join the piles of angular fragments skirting the bottom of the cliffs. The rivulets unite into streams and join the waters flowing from the glacier. Together, they tumble and eddy down the valley as a young and violent river.

This water, in global terms, is a rare liquid. Ninety-seven per cent of all the water on earth is salty. This, although it carries many rock particles in suspension, is chemically very pure. As it fell from the clouds through the atmosphere, it absorbed some carbon dioxide and oxygen, but very little else and so far it has had little chance to dissolve minerals from the newly exposed and largely unweathered rocks it has crossed. But gradually, as it rushes on, it collects organic particles from the mountain plants growing between the boulders near its margins, and eventually it acquires just enough dissolved nutrients to support animal life.

Blackfly larvae

Any creature that tries to make a permanent home in these rushing waters has to develop a method of preventing itself from being swept away. The larvae of the big hump-backed bloodsucker, the blackfly, attach themselves to the stones with a tiny circlet of hooks at their hinder ends and allow their legless, worm-like bodies to trail in the current. Occasionally, one moves farther into the river, curving down to grip a pebble with a small sucker at its front end, and then looping its body to get a new grip with its main hooks. Should it, during this manoeuvre, lose its hold, it can retrieve itself. It has spun a safety line of silk and fixed it to a pebble, so it is able to haul itself back to its original station. The swiftness of the stream, though it does cause problems, also brings one advantage. It ensures that, although the water contains relatively small numbers of edible particles, they do at least pass by with considerable frequency. All the blackfly larva has to do is to catch them. This it does with a pair of feathery fan-like structures on either side of its mouth. The larva pulls them down alternately and brushes off their catch with a pair of hairy mandibles. Before releasing each fan, it coats it with mucus from glands beside its mouth so that tiny particles, that otherwise would have passed between the filaments, stick to them.

Many species of caddis fly larvae live in fresh water. Lower down, in less turbulent rivers and in the still waters of lakes, they construct tubes from twigs or sand grains and move gently across the bottom, grazing on leaves and algae; but up here, where little plant food is to be had, the caddis larvae are hunters, trapping their prey with nets. One species spins a funnel of silk on the underside of a stone and lives inside it, grabbing other insect larvae or small crustaceans that might be swept past. Another makes a tubular net as much as 5 centimetres long, but with a mesh so fine that it can trap particles of microscopic size. This larva lives inside its net, periodically sweeping the inner surface with a bristly moustache on its upper lip. A third constructs an oval frame of silken ropes between the pebbles and then, crouching in front of it, swings its head in a figure-of-eight motion and weaves a fine net. The process takes no more than seven or eight minutes. If a large particle rips it, the larva quickly undertakes repairs. As the insect grows and becomes stronger, it ventures farther out into the stream, builds bigger and coarser nets and fishes for larger prey. With devices such as these, the larvae of caddis fly and a whole range of other insects – beetles and gnats, mayflies and midges – manage to colonise the mountain torrents. By doing so, they make it possible for other bigger creatures to live there as well.

If you walk down a high valley in the Andes, you may be lucky enough to see, perching on a boulder in the middle of the river, surrounded on all sides by a maelstrom of white water, a pair of most beautiful ducks. One, the male, has a white head streaked with black, a sharp cherry-red bill and a grey body. His mate has a grey head and reddish cheeks and breast. They are torrent ducks. The striking difference between their plumage is not assumed for the breeding season only, as is the case with many ducks, but exists all the year round. Suddenly, one of them will dive into the water and disappear. It is beneath the surface, facing upstream, bracing itself with its long stiff-quilled tail against a boulder, using small horny spurs on the wrists of its wings to give it

Torrent ducks, Chile

purchase, picking between the stones with its slender, slightly rubbery bill, collecting the larvae. After a minute or so, it bobs up and, without any apparent difficulty, regains its boulder for a few minutes' rest. For maybe half an hour the pair will work their way upstream from boulder to boulder, swimming with powerful strokes of their large webbed feet, judging to perfection the pace of the eddies and the rapids, landing on occasion on half-submerged stones with the torrent swirling around their legs, yet somehow nonchalantly retaining their foothold. Each pair has its own exclusive stretch of the river. When they reach the upper frontier of their territory they abruptly abandon themselves to the current against which they have been fighting so valiantly, and go racing back to their original position, bobbing and bouncing in the white water. Only very rarely do they leave the water over which they have such mastery and take to the air.

Torrent ducks live in high valleys along the entire length of the Andes, from Chile to Peru. In the north, they share the rivers with a bird of a very different ancestry but remarkably similar skills, the dipper. It is the size of a thrush, related to the wrens, and lives not only in the Americas but in the mountain streams of Siberia, the Himalayas and right across Europe to the British Isles. It takes tadpoles, small molluscs, little fish and surface-living insects, but it, too, is skilled in collecting larvae underwater. Its technique, however, is slightly different from that of a torrent duck. Its feet are not webbed, so it cannot rely on them for a powerful forward drive as the ducks can. Instead, it beats its wings below water and swims down to the bottom. Once there, it walks forward into the current, stabilising itself with quick wing-flicks, keeping its head down and its rump up, so that the current helps to counteract its natural buoyancy and keeps it pressed down towards the river bed. Many of the streams it inhabits in the northern part of its range, and in the high Himalayan valleys, are extremely cold, but dippers have very dense plumage and keep their feathers well waterproofed with oil from unusually large preen glands.

The powerful mountain rivers continue the work of destruction started at higher altitudes by ice and frost. During the dry season they may appear to be little more than gentle trickles, gurgling from one shallow pool to another, but you can judge how great their power becomes when they are in spate by looking at the boulders through which they wander. None has sharp edges like the fragments split by frost from the cliffs lying on the slopes above. All are rounded and smooth. Some may be gigantic, weighing many tons and even crowned on their summits by plants, showing that they have not moved for many years. Yet their smooth shape makes it clear that, in exceptional years when heavy rains have so swollen the river that the whole of the valley floor is filled with a brown roaring flood, these huge blocks have thundered down the river bed, smashing everything in their path.

As these young rivers make their way down from the mountains, they spout over piles of jammed boulders, glisten in lacy veils down steep rock faces and rush, in torrents of white water, through chains of rapids. If they have risen on the side wall of a steep valley or are flowing across a plateau, they may make a stupendous leap. In the

Dipper hunting under water, Britain

south of Venezuela, one river shoots over the edge of a sandstone tableland and falls clear for over 1000 metres as the Angel Falls, the tallest in the world. So high is it that, except in the wettest season, most of its waters are blown away as spray before they reach the ground.

Throughout this long eventful journey to lower lands, the river water becomes steadily richer. The stretches of mosses and cotton grass, heathers and sedges that drape the hillsides contribute their rotting leaves, turning the waters brown. The prolonged weathering of the rock faces and the corrosive effect of lichens and other plants change the minerals into chemical compounds that are soluble. Rock fragments, swirled round innumerable pot-holes and over many rapids, are reduced to tiny particles and carpet the river bed with stretches of sand and mud.

Now a great variety of flowering plants can put down roots in the river. The tug of the current is still strong and can threaten to tear them away. Many reduce that danger by producing underwater leaves that are divided into tassels, and only develop their large broad leaves above the surface where they cause no drag. The water is now much warmer and as a consequence it contains much less dissolved oxygen than it did when it was close to freezing point higher up the valley, but this impoverishment is, to a large extent, offset by the activities of the plants whose underwater leaves expel tiny bubbles of the gas as by-products of their photosynthesis.

Since the river is now warm, oxygenated and nutrient-rich, it can provide a wide variety of foods for fish – algae and plant leaves to be grazed; insect larvae, aquatic worms and small crustaceans to be collected; swarms of microscopic single-celled animals to be gathered by fish fry; and small fish to be caught and swallowed by bigger ones. But the unremitting flow of the waters causes problems to the fish as it does to smaller organisms.

Some, like brook trout, deal with the difficulty quite straightforwardly. They swim unceasingly. With beats of their tails, they exactly match the speed of the water which may be flowing past them at a rate of a metre a second. They maintain their position in a chosen pool where the feeding may be particularly good with ease and are so well within their strength that, should they be alarmed, they have no difficulty in suddenly thrashing their tails and shooting upstream to a new position.

Other fish, like the miller's thumb, manage to escape the pull of the water by sheltering between the stones of the river bed. In tropical streams members of two separate and unrelated families, the catfish and the loaches, have converted pairs of fins on their undersides into suckers. With these they can gain a firm grip on a rock. One catfish in the Andes and one loach in Borneo, however, have separately developed another method. Instead of anchoring themselves with a sucker, they have grown large fleshy lips and hang on with their mouths. This technique has one obvious drawback. They cannot, as most fish do, take in through their mouths the oxygenated water they need for their gills. Both fish have evolved the same solution to the problem – a strip of skin that stretches across the middle of the gill covers. They take in water through the top section and, having passed it over the gills, expel it through the lower one.

Angel Falls, Venezuela

Fish, like many other groups of animals, have alternative breeding strategies. Some take no care whatsoever of their eggs, but produce them in such vast numbers that a few will almost certainly survive. A female cod, for example, may release in a single spawning six and a half million eggs. Others, on the other hand, lay only a hundred or so eggs but invest a great deal of their time and energy in protecting them and guarding the fry.

The existence of a strong current flowing continuously in one direction, as it does in a river, has an obvious bearing on the relative advantages of these two techniques. It might seem that for a river fish to follow the first and abandon its eggs, as the cod does in the sea, would be totally impractical, for the helpless young would be swept away and then be faced with an almost impossible journey up the river if they were to return to their parent's birth place. Yet this is what both salmon and their close cousins, lake trout, do. The females lay their eggs in shallow scoops in the gravel and cover them with sand so that they are beneath the grasp of the current. One female may produce as many as 14,000 eggs. There they remain throughout the winter. When they hatch the following spring, the fry feed for a few weeks but eventually travel down the river over waterfalls and rapids. On reaching a lake, the trout remain there in its still waters. The young salmon, however, continue right down to the sea. When both have fed and become mature, they assemble in shoals and drive their way back up the river they descended, selecting almost infallibly the precise mixture of dissolved minerals and organic substances that characterise the water in which they hatched, until at last they regain their ancestral tributary. There they spawn. Many species now die. Others will return once more down the rivers and restore their strength in calmer waters before tackling the journey all over again the following year.

The arduous voyages of the salmon are not undertaken by many river fish. Most adopt the second strategy and shield their young from the river currents. The little miller's thumb lays its eggs in crevices in the rocks or even, on occasion, in empty mussel shells. The male maintains guard over them, valiantly attacking any other creature that approaches. The bitterling, another European fish, deposits its eggs not in an empty shell but in one that is still occupied by the living mussel. At breeding time, the female bitterling, only 6 or 7 centimetres long, extrudes a tubular ovipositor that is almost as long as she is. She carefully inserts it into the siphon through which the mussel expels water. She then lays a hundred or so eggs within the mussel's mantle cavity. While this is going on, the male hovers alongside. As his mate finishes laying, he discharges his milt which is caught by the current of water drawn in by the mussel and swept down its siphon on to the eggs inside. Thereafter, the fertilised eggs are kept well oxygenated by the steady flow of water that the mussel maintains through its shell for its own purposes. When the young bitterling hatch, they do not hurry to abandon their living sanctuary, but grip the soft flesh of the mussel's mantle with small horny outgrowths and feed and grow before they eventually release their hold and are swept through the mussel's exhalant siphon into the outside world.

It must be added that the mussel takes equal advantage of the bitterling. It breeds at the same time as the bitterling spawns and its tiny larvae are carried out from the shell

Mouth-brooding cichlid with young

over the adult bitterling and attach themselves to its gills and fins. There they remain until they are ready to settle down to adult life on the river bed.

In the Amazon, one small fish, the splashing characin, places its eggs out of reach of all waterborne dangers. Its solution requires the most gymnastic of breeding feats. Male and female lock fins and leap together out of the water on to the underside of a leaf overhanging the stream. For a few seconds, they hang on, held by specially long ventral fins which cling to the leaf, while they deposit a small clump of fertilised eggs. Then they drop back. For the next few days, the male patrols the water beneath, regularly splashing the leaf with his tail to make sure that the eggs do not dry out.

One family of freshwater fish, the cichlids, not only protect their eggs, but extend their care to the young. Over a thousand different species of them live in lakes and rivers all over Africa and South America. Some species deposit their eggs in depressions that they energetically dig in the gravel. Others deposit their sticky eggs on meticulously cleaned leaves or rocks, the female using her ovipositor to place them in neat rows with all the precision of an expert pastrycook icing a cake. As she does so, the male swims beside her, fins distended and tremulous and in full breeding colour, discharging his milt over the eggs.

Those cichlids that have the most straightforward ways of caring for their offspring now hover above the eggs, fanning them with their fins to ensure that oxygenated water keeps flowing over them. Other fish that approach are threatened with lowered throat and expanded gill covers, and even attacked and bitten. When the young hatch, many species excavate fresh nurseries in the gravel, pick the young up in their mouths and carry them across to their new quarters, carefully rolling them around with a chewing motion of their jaws so that they are given a good clean. As the fry grow and become mobile, the parents swim protectively beside them, taking up laggards at the back of the shoal into their mouths and spitting them out in a jet of water so that they are squirted right to the front.

Many cichlid species are even more solicitous parents. The mouth-breeders do not risk leaving their eggs in a nest. Immediately after they have been fertilised, one of the parents takes the whole spawning into its mouth and keeps them there for ten days or so. Throughout this time, the adult is unable to feed. It gently moves its jaw up and down so that the developing eggs are kept clean and free of bacterial infection. Even after they have hatched, the young stay within. Eventually, the parent spits them out, but if danger threatens it will, by depressing its jaw and throat, suck them all back inside. Even a week after hatching, the young continue to take refuge there. Sometimes they return in response to signals from their parent; sometimes they do so entirely of their own accord, nibbling the parental lips to seek entry.

Several African mouth-breeders have elaborated even this complex behaviour. The female, having laid her eggs, gathers them up before they are fertilised. The male, displaying nearby, has on his anal fin a row of yellow spots, outlined in black, of almost exactly the same size and colour as the eggs. Having collected all her actual eggs, the female, seeing these similar-looking objects on her partner's fin, swims across and

Discus with young

opens her mouth as if to gather them up, whereupon the male releases his milt so that the eggs are fertilised within the female's mouth.

Another cichlid, the discus fish, provides its young with a special food. The fish, as its name suggests, is a disc-shaped creature growing to 15 centimetres in diameter. Its olive-green flanks are magnificently patterned with iridescent stripes of red, green or brilliant blue. The female lays her eggs on stones or leaves. When they hatch, both parents carefully transfer them to other leaves where they hang by thin threads. The adults then develop a covering of mucus on their bodies. It exudes from their flanks and even veils their eyes. The youngsters detach themselves from the leaves, wriggle across and for the next few days graze over their parents, eating the protein-rich slime.

The greatest protection that any animal can give its offspring is to allow the eggs to hatch inside the female's body and to stay there until such time as they have passed the first stages of development when they are at their most helpless and vulnerable. The technique is employed by all mammals, except the marsupials, and may be counted as one of the characteristics that have contributed to the group's success. But fish were using a similar technique long, long before the mammals existed. In the sea, sharks and rays still reproduce in this way and many families of freshwater fish have also come to do so. The little guppy is only one member of a large family of live-bearers that swarm in tropical rivers and lakes. The male's anal fins are modified to form a small movable tube, called a gonopodium, through which bullets of sperm are shot into the female's aperture. The male fusses around the much bigger female, assessing her readiness for breeding, and if satisfied on that count, taking aim. Then he darts in and momentarily brings his gonopodium into contact. One successful shot is sufficient to fertilise several batches of eggs which hatch inside the female. The young within her become visible as a dark triangular patch at the back of her body. Eventually, they emerge, one at a time and sufficiently well-formed to be able to swim swiftly to shelter from danger among the plants.

One of the several species of top-minnows, which live in the rivers of southern Brazil develops this sexual apparatus in a very unusual way. The gonopodium is formed not only from fin rays but from skin and, as a result of this construction, is not as mobile across such a wide range as that of the male guppy. Indeed, a male top-minnow can point his gonopodium to one side only. Some do so to the right, some to the left. The female top-minnow's aperture is also asymmetric in a similar fashion. So left-shooting males can only mate with right-targeted females.

These large and varied populations of river fish inevitably attract predators. The fish themselves have produced some of the most ferocious – the piranhas of South American rivers. These are, for the most part, small creatures – the largest of the many species is not more than 60 centimetres long – but they have formidable teeth, triangular and so sharp that the Amazonian Indians use them as scissors. The piranha normally prey on other fish, usually wounded or diseased individuals, but they will also attack much larger creatures – tapirs, capybaras, horses – that might be swimming in the rivers. They attack in shoals. As they feed on a body, dead or alive, they become more and

Piranhas and other fish, South America

more frenzied as more blood flows into the water and they compete with one another to strip the last mouthfuls of flesh from the bones. Frightening though such an attack may be, the danger piranhas pose to human beings is often much exaggerated. They seldom attack unless an open wound is tainting the water and they do not lurk near rapids where a traveller is most likely to need to wade or to be tipped out of his canoe.

The river fish are attacked by other kinds of hunters. Turtles lie in wait for them on the bottom. They are not swift swimmers and catch their prey by stealth. The matamata, a South American species, disguises itself with tatters of skin that hang from folds and dewlaps of its head and neck. Its shell, too, is uneven and often sprouts a fur of algae. When the animal lies on the bottom among rotting leaves and twigs, as it often does, it is virtually invisible. If a fish strays within range, the turtle suddenly gapes – and the fish is engulfed. The alligator snapper turtle, one of the largest of all freshwater species, growing up to 75 centimetres long, is a more active fisherman. It has a small projection on the floor of its mouth which ends in a bright red worm-like filament. It rests with its jaws agape and, every now and then, twitches its little red bait. If a fish comes to collect it, the turtle simply shuts its mouth and swallows.

Crocodiles and their American cousins, the caiman and alligators, pursue fish when they are young, but change their diet to carrion when they become adult. In India, however, lives another member of the family, the gavial, that eats nothing but fish throughout its life. It has long thin jaws which are much easier to clap together underwater than the broad jaws of a crocodile and it fishes with a sideways sweep of its head. It is an enormous reptile, said to grow to 6 metres long, but the muscles needed to grasp a fish are weaker than those used by a crocodile to rip the leg from the carcass of an antelope, and gavials have a relatively feeble bite. They have never been recorded as attacking human beings.

Now that the river has reached its middle stretches, it has abandoned the energetic leaps, the speed and the erratic course of its youth. No longer does it grind and tear at the land through which it passes. Middle age has come upon it. Its slower broader waters may still be turbid, but it is more likely to be depositing sediment than picking it up. Mud washed into it from the forests and grasslands on its banks make its waters more fertile than they were. Thickets of trailing plants sway back and forth in its gentle current. Rushes and reeds line its margins and clog its backwaters, and land animals of all kinds come to drink its water and plunder its populations.

The weasel family, all ferocious and skilful hunters, includes one member that has become a fish-eating specialist, with webbed feet, closable ears and water-resistant fur – the otter. It pursues fish underwater with speed and a weaving sinuous persistence that few fish can dodge. Sometimes an otter will slap the water with its tail and drive shoals of fish into shallow pools where the panicked fish are caught with even greater ease.

On the banks above perch kingfishers. Some can hover as skilfully as hawks, hanging in the air on beating wings. When one sees an incautious fish approach the surface, it plunges headlong, seizes the fish with its sharp bill and returns to its stance. There it

Alligator snapper turtle

beats its catch several times on the perch to stun or kill it, and juggles it so that when, with a final toss, the bird swallows the fish, it goes down head-first with the spines on its fins pointing backwards and not snagging in the bird's throat.

At night, in southeast Asia and Africa, owls come down to the river to fish. Their legs are bare of feathers, so they can strike cleanly through the water, and they have sharp-edged spiny scales on the underside of their feet which enable the birds to grasp their wriggling slippery prey more firmly. Their flight and swoop seem unexpectedly noisy to anyone who has watched woodland owls. Those birds have wings that are specially silenced with downy fringes on the flight feathers. But fishing owls have no need for such silencers, for fish, unlike voles and mice, are not very sensitive to airborne noises.

No owls fish in the Americas. The talons that sweep through the surface waters there belong not to birds but to bats. It seems that there is not room for two creatures to practise this kind of fishing and in the New World the bats were the first to develop the technique and have retained the night-time fishing rights ever since.

Other land animals come to the rivers to crop the water plants. In Europe, water voles – often inaccurately called water rats – with chubby faces and furry tails, busy themselves mowing the grass along the banks and felling the reeds. Though they are very competent swimmers and divers, they do not have any specific physical adaptations to aid them in the water. The beaver, on the other hand, which also once lived in considerable numbers in Europe and is still abundant in parts of North America, is a very well-equipped swimmer indeed. Its hind feet are webbed, its fur dense and water-repelling, its ears and nostrils closable, and its tail flat, broad and naked so that it serves as an excellent oar with which it paddles. Beavers dig for lily roots and chew bulrushes, but they find their main food, not in the river, but on its banks, where they strip the bark and chew the twigs and leaves of deciduous trees such as aspen, birch and willow. They also gnaw through and fell trees with trunks as much as half a metre in diameter. These they drag down to a stretch of the river where the water is shallow. They pile mud, boulders, more branches, tree trunks and piles of vegetation on them until they have constructed a barrier right across the river, damming its flow and building up a lake of considerable size. On its shore, these indefatigable animals build their lodge, a large dome-shaped structure, with one or more underwater entrances, in which the whole family lives. This lake, created with such labour, serves as a larder. The beavers drag down branches of trees and bushes and sink them in its waters so that in winter, when the land is covered with snow and the lake plated with ice, green bark can still be fished up from the water and eaten. They can get beneath the thickest ice on the lake through the unfrozen entrances within the lodge. The lake also gives them great security, for as long as they keep the dam in repair and the water level does not fall, those entrances will not be exposed to the outside world and their lodge will be burglar-proof.

The biggest river-dweller of all, the hippopotamus of Africa, also uses the river for protection rather than as a pasture. Herds of them are a common sight during the day, lounging in the rivers, grunting, gaping and occasionally squabbling. The water buoys

Iguazu Falls, South America

up their immense cumbrous bodies so that they move with ease through it, tiptoeing along the bottom with the lightest of touches of their feet. Because this is when and how we usually see them, we tend to regard them as habitual river animals, but their most active periods are spent on land, and at night. In the late evenings, they plod up the river banks, often along tracks that the herd has used for generations, and crop the grass, each animal eating as much as 20 kilos in a single night. Before dawn, they return to the river where nothing, not even a crocodile, is big enough to attack them. Their movement back and forth between the land and the water is of great importance to river creatures, for the hippopotamus habitually defecates in water. So every day, the animals deliver to the river a load of nutrients synthesised by the land plants, and shoals of fish are always swimming around their rear ends waiting to consume the next instalment.

As the river continues down towards the sea, its course may be impeded by a band of rather harder rock against which its cutting tools of sand and pebbles can make little impression. As the gradient slackens, the river broadens until it reaches the farther edge of the hard layer and spills over it to resume its erosion below. This produces a cliff across the river's course and over it, a waterfall. Such is the origin of most of the great falls of the world – the Victoria Falls on the Zambezi, the Iguazu Falls on a tributary of the Parana in South America, and Niagara on the river running between two of the Great Lakes in North America.

None of these can compare in height with the dizzy leap of the Angel Falls, but in terms of their width and the volume of water that passes over them, they are incomparably bigger. They may not be able to erode the upper surface of the barrier which creates them, but they are able to attack it from below. The water pouring over the lip of the fall pounds down on the softer rocks at the foot, wearing them away and undercutting the hard layer until blocks split off from the edge and tumble down the face of the fall. So these huge cascades steadily work their way along the river, leaving a deep gorge behind them downstream. Niagara, at the moment, is moving at a rate of over a metre a year.

These gigantic waterfalls create their own microclimates. The mass of tumbling water displaces gales of air which blow up the walls of the gorge beside the fall, drenching them in spray. At the Victoria Falls, this produces a miniature rain forest in sharp contrast to the baked savannahs all around, where orchids, palms and ferns all flourish and where, mingling with the roar of the water, you can hear the calls of frogs and the whine of insects.

At Iguazu, the rock behind the curtain of falling water is used as a sanctuary by swifts. During the day, they hunt high in the sky, almost beyond sight, hawking for insects. As evening approaches, they assemble in immense flocks, still at great altitudes, until just before sunset, when they begin to stream down at high speed. They dive straight at the wall of water. Just before they hit it, they fold their wings and their velocity carries them straight through to the rock face behind. With an upward swoop, they bring their feet forward and clutch the rock and there they hang, some in a dry spot, some with a runnel of water spouting over them, but apparently enjoying the

River meandering through jungle, Brazil

bathe, preening themselves and occasionally drinking. To human eyes, they seem to court great danger for the disproportionately small reward of a perch, but so great is their aerial skill, and so unfailingly do they succeed in diving through the water, that one can only conclude that, for them, there is no risk whatever in reaching their impregnable roost.

The rivers are now coming to the end of their journeys. They are old – fat and slow-moving. They still carry some sediment, but do so erratically, picking some up here, dropping it there. As they sweep round a bend, the water on the outside, having to travel much farther, will necessarily move faster than that on the inside. So the sediment remains in suspension on the outside of the bend and cuts away at the bank, while on the inside it falls to form banks of shingle and mud. So the ageing river gradually works its way sideways across the plains. Sometimes it swings and meanders in such an exaggerated way that one bend approaches another until finally the neck of land separating the two becomes so narrow that it collapses. Then the river takes the shorter course and a curving length of its bed becomes isolated as a lake.

There the waters are still. The factor that governed so many of the habits and structures of the river creatures, the eternal tug of its current, has disappeared. Life can take on new forms. Plants no longer hug the banks or fix themselves to rocks. Now they can allow their leaves to float to the surface and catch the maximum amount of light. Water lilies, rooted in the thick oozy sediment on the bottom, send up their shoots and unfurl circular pads. The biggest of all, the famous Victoria lily of the Amazon, does this so aggressively that it ousts other plants from its part of the lake. Its immense leaves, strengthened by stout air-filled ribs and armed beneath with spines, have high up-turned rims. As they expand to their full diameter of 2 metres, these rims advance across the surface, pushing aside all other floating plants to claim the space for themselves. Their flowers, as big as soup plates, are white when they first open. They produce a scent that is particularly attractive to beetles, which come lumbering through the air to feed on special sugar-laden outgrowths in the centre of the flower. A fully developed bloom may attract as many as forty of them. Most bring with them loads of pollen which they have gathered from other flowers and which they now spread on the female parts of the blossom. In the afternoon, the lily's petals slowly shut and the feasting insects are taken prisoner. They remain trapped until the following day when the petals open once more. By now the beetles are thoroughly dusted with a new coat of pollen and they fly away, taking it with them, to feed on another plant; and the bloom, having been fertilised, slowly turns purple and dies.

Across these immense leaves walk elegant plover-sized birds, jacanas or lily trotters. Their toes and nails are hugely elongated so that the small weight of the bird is spread over a considerable area of floating leaves. Jacanas do not restrict their trotting to lilies but roam widely over carpets of much smaller floating plants. They even nest on the water by building a raft of buoyant water leaves and anchoring it among the reeds. They eat a certain amount of vegetation but they spend much of their time hunting for small insects running among the floating plants and over the surface of the water.

Leaves of Victoria lily with jacana, Brazil

Water remains a liquid substance and not a mass of dispersed droplets because of a strong physical force, akin to magnetism, that attracts one water molecule to another. The molecules that form the surface of the water have only molecules of gas above to which they are not nearly so strongly attracted. Their forces, therefore, are concentrated instead on the water molecules below and alongside them. The exceptionally powerful bonds caused in this way give the water a kind of elastic skin which is sufficiently strong to support tiny insects. A whole population lives on this springy bouncy platform and exploits its extraordinary properties.

If an animal is to rely on this molecular skin for support then clearly it must not break it. That can be prevented with wax or oil, both of which physically repel water molecules. So pond skaters, which have wax-coated feet, are able to stand on the surface, their six tiny legs splayed wide, each creating a minute dimple on the surface film. Springtails no bigger than a pinhead have wax all over their bodies. They, however, are so small and so light that their problem is not how to prevent themselves breaking through the surface but how to stop themselves being blown off it. They hold on to the water with a tiny peg beneath their body which lacks wax and so pierces the film and is gripped by it. Their legs, too, are tipped with waxless claws which penetrate the film and give the animals traction.

The springtails feed on pollen grains and algal spores that settle on the water. Most of the other surface-living animals rely for food on the bodies of small insects that get blown here. These do not sink because of the buoyancy of the water beneath, but the water molecules in the surface skin bond with those that have soaked into their bodies so that the fallen insects are gripped by the surface tension. It is as though they have landed in glue. Their struggles to free themselves cause vibrations to ripple across the elastic surface. The water-walking hunters react quickly and hurry across. The first to arrive immediately drags the victim clear of the surface so that its struggles will no longer be detectable by others and the skater will have the meal all to itself. Swamp spiders sitting on the bank at the water's edge rest their front legs on the water and respond to the vibrations of the surface film in just the same way as their land-living relatives react to the movements of their webs. As one rushes out on its eight water-repelling feet to the source of the vibrations, it pays out a silken rope attached to its base on the bank with which it will haul itself and its prey back to land.

Whirligig beetles extract a different kind of information from ripples. They create them themselves by continually gyrating on the surface. They then monitor the returning ripples and are able from them to detect the presence of obstacles around them. Pond skaters are even more sophisticated in their reading of the ripples. They shake their bodies like frenzied gymnasts to vibrate the surface film with special and characteristic frequencies to tell other pond skaters that they are ready to mate.

Perhaps the most spectacular use of the surface tension film is that made by the camphor beetle. It lives mostly on land at the margins of the water, but should one fall on to the surface of the water, it escapes from the skaters and the spiders by producing a special chemical from the tip of its abdomen that reduces the attraction between water

Pond skater

Water spider catching stickleback

molecules. With surface tension no longer holding its rear but still pulling at its front legs, the beetle shoots across the surface as though it were powered by a tiny outboard motor. It can even steer by flexing its abdomen from side to side and it usually manages to regain the shore and safety by a turn of unmatchable speed.

The lakes created by cut-offs from a meandering river are relatively small. Bigger ones come into existence in other ways. Some have formed in valleys dammed by an avalanche, by ridges of rock debris pushed up by now-vanished glaciers, or by the engineering skills of human beings. Lake Baikal in Central Asia and the lakes of East Africa have accumulated in immense cracks that have developed in their continents as a result of major crustal movements. The Great Lakes of North America lie in a basin created during the Ice Age when an ice cap blanketed most of the continent. Not only did glaciers flowing from it gouge out deep basins in the valleys down which they flowed, but the whole area was depressed into a bowl by the weight of the ice pressing the continent down into the plastic basaltic layers beneath. Since then, the ice melted relatively quickly, but the continent has not yet returned to its proper level.

Around the margins of the great lakes, in shallow bays among the rushes, life can be very like that in smaller stretches of fresh water. Dragonflies and damsel flies, midges and mosquitoes breed among the vegetation; snails and mussels live in the mud; pike and piranha hunt; carp and cichlids crop the vegetation. But where the lake floor falls to great depths, then conditions change radically.

Baikal is the deepest lake in the world, its bottom in places lying 1.5 kilometres below its surface. This is not exceptionally deep compared to the ocean, but whereas currents run across much of the ocean floor, few disturb the enclosed world of the great fresh-water lakes. River water, flowing into the lakes, is comparatively warm and so floats over the top of the cold deep water. Occasionally major storms may so disturb the surface layers that the waters are stirred to some considerable depth, but for most of the time, the lower parts of the great lake are near to freezing, black, poor in oxygen and, in spite of many a legendary monster, largely lifeless.

These lakes do not, however, lack for biological distinction. Since they are isolated bodies of water, animal communities once established in them receive few new recruits. The only way that wandering water-living creatures can reach them is along the rivers. To travel upstream to them would require swimming against the current, across other smaller lakes and up waterfalls. Few do so and most of the inhabitants of the great lakes are descended from species that lived in their headwaters. Slight genetic changes that may arise in individuals among these small communities are not swamped, as they may be in bigger breeding populations, and as a consequence are the more easily preserved. So lake animals tend to develop into their own characteristic species. Lake Tanganyika is about one and a half million years old and contains 130 species of cichlids and 50 of other fish that are now unique. So are many of its shrimps and mussels. Lake Baikal has, perhaps even more remarkable inhabitants. It contains 1200 species of animals and 500 plants, over 80 per cent of which exist nowhere else. There are huge flatworms, red and orange, striped and speckled; a bullhead fish which lives on the floor a kilometre down; and molluscs which, since the lake waters are not as rich in calcium salts as the sea, produce shells that are much thinner than those of their marine relatives. The lake also has its own unique mammal, a seal. It is very similar to and almost certainly descended from the ringed seal that lives in the Arctic. But the lake is over 2000 kilometres from the Arctic Ocean and to reach it by river would necessitate the crossing of innumerable rapids and waterfalls which would seem to be beyond the abilities of any seal. It is possible, however, that the seals first ascended the rivers to the lake during the Ice Age, when the journey may have been much shorter and easier. Today, the Baikal seal is not only the sole member of its family living in freshwater but is considerably smaller than any other.

In geological terms, lakes are transient features of the earth's surface. The cut-off meanders may disappear within decades. The bigger lakes may last for thousands of years but even they are shrinking. The rivers, as they enter the still lake, drop their loads of sediment to form deltas that slowly extend across the lake filling its deeps. The water around the margins becomes shallower as sediment is washed into it from the surrounding land by streams. As the bottom approaches the light, plants get root and clog the waters still further with their stems and regular deposits of rotting leaves and roots. So the lake becomes first a swamp, then a marsh, and ultimately a fertile meadowland through which the river that first filled the lake still wanders.

On the plains that lead down to the coast, the rivers play out the last acts of their old age. The slope is now so gentle, the movement of the waters so slow, that the rivers shed all but their finest particles. Sandbanks and mudbanks split their channels repeatedly, so the rivers braid into a maze of branches.

Hundreds of kilometres away, in the high mountains around their source, storms pour water into their tributaries. Days later, the aged rivers suddenly swell, rise above their banks and flood the plains, depositing layers of fine mud. These sudden and regular immersions can create green fields in a desert, as the Nile does in Egypt. In temperate lands, they produce plains of great fertility where crops grow in abundance, as cotton does on the delta of the Mississippi. The flood plain of the Amazon extends over much of northern Brazil. Most of it is still jungle-covered and the river's beneficiaries are the huge trees. When the floods come, the river's fish swim out between the trunks over the submerged land to find food. Many gather the fruits that drop from the branches above. This diet is not an optional one, adopted sporadically when the chance presents itself. It is the main banquet of the year during which the fish will put on fat that will sustain them through the leaner season when they are confined between the river's banks. Catfish have developed particularly large mouths that enable them to tackle fruit. Species of piranha have evolved that do not eat flesh but feed almost entirely on such fruits. Some species of characin fish have large crushing molar teeth and such powerful jaw muscles that they can even crack Brazil nuts. The trees' seeds, however, are not destroyed by the fishes' digestive juices. They survive and are excreted elsewhere in the shallow flood waters. Such Amazon trees, it seems, rely on fish to distribute their seeds just as trees in the jungle elsewhere rely on birds. Here, too, many of these fish spawn, for these waters, rich with rotting vegetation, produce a great deal of microscopic organisms on which their young can feed.

Now at last, the rivers are approaching the sea. For some, the journey from their source has been only a few miles. For others it has taken them halfway across a continent and lasted for months. The Amazon is the greatest of all the world's rivers, over 6000 kilometres long. At any one time, two-thirds of the world's fresh water is flowing between its banks. At its mouth, it is 300 kilometres across, a maze of channels and islands just one of which is, by itself, bigger than the whole of Switzerland. And this giant river maintains its identity even after it has left the coast. In 1499, a Spanish sea-captain sailing down the east coast of South America well beyond the sight of land suddenly became aware that the water he was crossing was not salty but fresh. He turned west and became the first European to see this gigantic river. Not until its waters are 180 kilometres beyond the edge of the continent do they lose their identity and mingle at last with the salt waters of the ocean.

NINE

THE MARGINS OF THE LAND

All great rivers, the Amazon and the Zambezi, the Hudson and the Thames, as well as thousands of smaller ones are turbid with sediment by the time they reach their estuaries. Even the clearest of their waters are loaded with microscopic particles of minerals and decayed organic matter. When these mingle with the dissolved salts of sea-water, they clump together and drift to the bottom to form great banks of mud.

Estuarine mud has a fineness, a stickiness and a smelliness all of its own. If you step into it, it clings so tenaciously that it can suck the boot from your foot. It is so fine-grained that air cannot diffuse through it, and the gases produced by the decomposition of the organic debris within it stay trapped, until your footsteps release them and produce a whiff of rotten eggs.

Twice a day, the character of the water flowing over the mud flats changes radically. When the tide goes out, and especially when the rivers are swollen with rain, fresh water will predominate; when the tide comes in, the water in the estuary may become as salty as the sea; and twice a day, too, a great part of the mud will not be covered by water at all, but exposed to the air. Clearly, organisms that live in such a place must be able to withstand a great range of chemical and physical conditions. But the rewards for doing so are huge, for food is delivered to the estuary every day from both the sea and the land, and its waters are potentially more nutritious than almost any others, salt or fresh. So those few creatures that can survive here flourish in immense numbers.

In the upper part of the estuary, where the water is only mildly brackish, hair-thin sludge worms live with their heads buried in the surface of the mud, eating their way through it, waving their tails in the water above to bring down a current of oxygen-bearing water. A quarter of a million of them can live in just one square metre of mud, covering its surface with a fine red fur. Farther down towards the sea, where the water is a little saltier, vast numbers of tiny shrimps, a centimetre long, build burrows for themselves and sit snatching at passing particles with their hooked antennae. Spire shells, each scarcely bigger than a grain of wheat, work their way through the creamy layer of recently deposited mud to extract such goodness as it contains. They flourish so greatly that 42,000 have been extracted from just one square metre.

Oystercatchers, Britain

A little nearer low-water mark, and especially where sand is mixed with mud, lugworms burrow. They too are eaters of mud, but they enrich it before they consume it. Each worm, about 40 centimetres long and as thick as a pencil, digs a U-shaped tube, lining and securing its wall with mucus. It fills one arm of this with loosely compacted sand grains. Then, gripping the sides of the tube with the bristles on its flanks, it moves up and down at the bottom of the tube like a piston in a pump, sucking water through the sand plug. The particles the water carries are trapped in the sand. After a time, the worm stops pumping and begins to eat the sand, digesting the edible bits and excreting the rest into the other arm of its tube. Every three-quarters of an hour or so it pushes this processed sand out of the top of the tube to form a cast. Cockles too lie buried in this region, just below the surface. They do not compete with the lugworms for the mud, but put up two short fleshy siphons to suck in particles directly from the water.

When the tide goes out, all these creatures have to stop feeding and take steps to prevent themselves being dried out. The mud around the spire shells is so little compacted that much of it is carried away by the draining waters and the tiny shells lie in concentrated layers, inches deep. Each of them seals the entrance to its shell with a small disc fixed to the end of its foot. The cockles clamp the two halves of their shell together to form a watertight seal. The lugworms simply withdraw into their tubes which are so deep they always remain waterlogged.

But there are dangers other than desiccation that now threaten them. All are vulnerable to attack from the air, and huge flocks of hungry birds descend on the estuary. The food each kind of bird harvests is determined to a considerable degree by the size and character of its beak. Tufted duck and pochard dabble through the mud extracting the sludge worms. Ringed plovers, with short sharp bills, feast on the spire shells, extracting each little fleshy coil with a flick of their bill. Knots and redshanks, whose bills are twice as long, probe into the top layer of mud for shrimps and small worms. Oystercatchers, with their stout scarlet beaks, catch the cockles. Some individuals prise the shells apart. Others habitually prefer smaller cockles with thinner shells and specialise in hammering them to pieces. Curlews and godwits, which have the longest beaks of all, probe down deep enough to reach the lugworms and drag them out of their burrows.

As the rivers bring down more and more sediment, so the mud flats slowly rise. A skin of green algae begins to form on them, binding the mud particles together. Once that happens, other plants can get a roothold. The mudbanks now begin to grow with increased speed, for the mud particles carried in by the lapping waves no longer swill back but are caught by the roots and stems of the plants. Eventually, their surface is sufficiently far above the water to be beyond the reach of all but the highest tides. So the banks are fixed and the creatures of the estuary have lost their territory to the animals of the land.

Around European shores, this pioneering reclamation of the land is carried out by glasswort, a small plant which, with its scale-like leaves and bloated translucent stems, has the appearance of a succulent water-conserving plant from a desert. The comparison is indeed an apt one. Flowering plants evolved on land and their chemical processes are

Glasswort, Britain

all based on fresh water. Sea water causes them great problems for since it contains a higher concentration of salts than their sap, water tends to flow out from their tissues through their roots rather than into them. So it becomes as important for plants in a salty environment to conserve their water as it is for a cactus in a desert to do so.

In tropical estuaries, the task of anchoring the mud is performed by mangroves. There are many species of them, some no bigger than large bushes, others tall trees 25 metres high. They come from several different plant families, but the requirements of living in brackish swamps has led them all to develop very similar characters.

Maintaining a hold on the glutinous shifting mud is a major problem for a plant as big as a tree. It cannot be done by sending down deep roots, for the warm mud only a few centimetres below the surface is without oxygen and corrosively acid. Instead, the mangroves develop their roots into a broad horizontal platform that sits, like a raft, on top of the mud. Some of the taller kinds gain added support by sending out, from points quite high up on their trunks, curving aerial roots which serve as struts. Roots must provide a tree with sustenance as well as stability and the shallow system of the mangroves is well suited to do this, for the nutrients the tree seeks lie not deep in the acid mud but on its surface, where they have been deposited by the tides.

Roots also provide trees with an escape route for the carbon dioxide produced by the living processes, as well as an entrance for the oxygen. Again, there is no oxygen to be had in the mud. The mangroves draw it directly from the air through small patches of spongy tissue that develop on their bark. These are placed on the aerial strut roots of those mangroves that possess them. Those that do not, produce the patches on large knee-like flanges that rise on their horizontal roots. The mangrove species that grows nearest the sea, where mud is being deposited most rapidly, produces lines of conical air-absorbing roots that grow not downwards like normal roots, but vertically upwards, so keeping pace with the mud and creating around the tree a carpet of sharp spikes that looks like some fantastic medieval defence system.

Salt causes difficulties for the mangroves, as it does for the glasswort. They, too, must conserve the water within their tissues, and they prevent evaporation from their leaves by the same devices used by desert plants – a thick waxy skin, and stomata placed at the bottom of little pits. They also have to prevent a high concentration of salt forming in their tissues which would seriously upset their chemistry. Some of them manage to exclude it from the brackish water they absorb with a special membrane covering their roots, as the glasswort also does. Others, lacking this ability, accept the dissolved salt into their roots, but then get rid of it before its concentration becomes dangerously high. This they manage to do either with special glands on their leaves that exude concentrated brine, or by transporting the salt in their sap, depositing it in their old leaves and, in due time, shedding those leaves and with them the unwanted salt.

As mud accumulates on the seaward edge of the swamp, the mangroves advance and claim it. They do so with special seeds, which germinate while they are still hanging on the branches, putting down a stout green shoot like a spear which, in some species, may be as much as 40 centimetres long. Some of these drop directly into the tangled roots

Air-absorbing roots of mangroves, Bangladesh

below and wedge there. The lower tip sprouts rootlets, the upper end a stem and leaves. Others, falling when the tide is in, float away. At first, in the brackish estuarine water, they hang vertically, but if they are carried by the falling tide out to sea, the greater buoyancy of the saltier water gives them more support so they tip over and float horizontally. The green cells in their skin are now able to photosynthesise, providing food for the young plant. The delicate bud at the end, from which leaves will eventually spring, is kept wet and cool and unscorched by the sun. In this position, the infant mangrove can remain alive for as long as a year, during which time it may drift for hundreds of miles. If eventually currents should carry it into another brackish estuary, then it reverts to its original erect position with its root pointing downwards. When the tip catches in the soft mud on a falling tide, it embeds and sprouts rootlets very rapidly; and a new mangrove tree is established.

Although a few clear channels may run through a mangrove swamp, the trees grow so thickly over most of it that you cannot force even a small boat between them. If you want to explore it, you have to go on foot when the tide is out. It is not a comfortable place. The dense arching aerial roots are not, for the most part, strong enough to bear your weight without bending, so your foot skids off them. Many are encrusted with shells whose sharp edges will slice your legs as you slip, or cut your hands as you grab a root to prevent yourself falling headlong. The smell of rotting hangs everywhere. Water trickles and drips from the roots. Clicks and plops sound in the heavy air as molluscs and crustaceans move in their holes, snap their claws and shut their shells. Mosquitoes whine around your head and stab your skin. The leafy branches above are so thick that not even the slightest breeze relieves the heat, and the air is so humid that sweat cascades from you. Even so, the swamp has an undeniable beauty. The water swilling through the roots casts a silvery shimmer on the underside of the leaves. The vaulting strut-roots that criss-cross one another, and the spikes and knees of the air-breathing roots projecting from the mud, create endless changing patterns. And everywhere, there are animals.

A whole army of different creatures is busy, gathering the new helping of food that the retreating tide has left behind. Small sea-snails, not unlike winkles, move slowly over the mud grazing on fragments of algae. Ghost crabs, 5 centimetres across, sprint over the mud, searching for organic refuse, keeping a sharp lookout for danger with eyes that are not placed on the tip of long stalks, but surround them, giving their owners 360° vision. Fiddler crabs emerge cautiously from their holes and begin to work through the surface layers. They pick up a tiny lump with their pincers and deliver it to the set of hair-fringed blades that scissor to and fro in front of their mouths. One set of spoon-shaped hairs retains a clump of sand grains, while another flicks off anything that might be nourishing into the mouth behind. The inedible grains accumulate at the bottom of the mouthparts where they are consolidated into a little pellet, picked off by a pincer, and then dumped as the crab moves forward a few steps to take another clawful.

Female fiddlers perform this operation with both pincers. The males, however, have to do it one-handedly, for while one of their pincers is like those of the female, the other

Ghost crab, Australia

is greatly enlarged and conspicuously coloured, pink, blue, purple or white. Its function is not that of a serving fork but of a signaller's flag. The male waves it at the female while at the same time performing gymnastics. The exact combination of choreography and semaphore varies with the species. Some stand on tiptoe and wave their claws in circles; others frantically swing them to and fro; still others hold their claws still and jump up and down. The messages, however, are all the same: the male is ready to mate. One of the females, recognising the particular gestures of her own species, will eventually respond by trotting towards a male and then following him down into his burrow and mating with him.

Crabs as a group originated in the sea, and most species still live there, where they breathe by passing oxygen-laden water through gill chambers within their shells. The fiddlers, however, must breathe out of water. They do so, quite simply, by retaining water within their gill chambers even though they themselves have emerged into the air. The oxygen in this small volume of water is, of course, soon exhausted, but the crab renews it by circulating the water through its mouthparts and beating it into a froth. Having absorbed more oxygen from the air, it is then returned to the gill chamber.

Fish too come out of water, wriggling across the mud between the mangroves. These are mudskippers. The biggest of them is about 20 centimetres long. They use the crabs' breathing technique of keeping their gill chambers full of water, though they have no way of circulating it to renew the oxygen, and regularly return to the water's edge to collect a new and refreshing mouthful. But they have one absorbent surface which the hard-shelled crabs lack – their skin. They take in much of their oxygen through this, just as a frog does, but if it is to serve this purpose, the skin has to be wet, so the mudskippers, as they move around, make swift sideways rolls to moisten their flanks.

If they want to move at speed, to catch a crab or escape danger, they curl their tails sideways, flick them, and shoot across the mud. Most of the time, however, they travel much more soberly by crutching themselves along with their two fore-fins. These are strengthened internally with bony struts, well-muscled, and equipped with a joint halfway down, so that the mudskipper appears to be heaving itself across the mud on its elbows. In some species a second pair of fins, further down the body on the underside, have become joined to form a suction cup with which the fish can cling to mangrove roots.

Mudskippers live in mangroves in many parts of the world. Each swamp usually contains three main kinds. One, the smallest, remains in the water for the longest time and only ventures out of it at low tide. Shoals of them wriggle through the liquid mud at the water's edge sifting it for tiny worms and crustaceans. The mid-tide area of the swamp belongs to a much bigger kind. This is a vegetarian which collects algae and other microscopic single-celled plants. It feeds alone and is strictly territorial, digging a hole for itself and patrolling the mud around it. It sometimes builds low mud ridges several metres long around its boundary, so keeping out its neighbours and, to some extent, preventing the mud from draining completely. Where the population is high, these territories abut one another and the mud flat becomes divided by the walls into

Mudskippers beside their pond, Malaysia

polygonal fields, each with its owner presiding over it like a bull in a paddock. A third kind of mudskipper occupies the highest part of the swamp. It is a carnivore and preys on small crabs. It too burrows, but it is not so proprietorial over its surrounding territory and shares its hunting range with its neighbours without dispute.

The mudskippers not only feed out of water but conduct their courtship there. Like most fish, they display by flexing and quivering their fins. Since both their sets of paired fins are dedicated to locomotion, they have to display with the two long fins that run along their spine. Normally they lie flat, but when courtship starts the male erects them, revealing them to be brilliantly coloured. This, in itself, is not enough to attract a mate from any distance away, for a mud flat is very flat indeed and a small fish sitting on it will only be seen by its immediate neighbours. So a male mudskipper, intent on making his attractions known to the widest audience, flips his tail and leaps vertically into the air, banners erect.

The low-tide species, as far as is known, does not care for its young when they hatch. The tiny creatures are swept away and join the community of larvae and fry that floats suspended in the surface waters of the sea. The vast majority of them will be eaten or carried into parts of the sea away from the swamps, where they cannot survive.

The mid-tide species, however, gives its offspring better protection. The male digs a burrow in the middle of the walled territory and constructs a circular rampart around the mouth of it. The mud here is so close to the level of permanent ground-water that this creates a little walled pond. The male then lounges on the wall of his swimming pool where the female joins him. Mating takes place in the privacy of the burrow at the bottom of the pool. There, too, the eggs are laid and the young will remain, even when the tide is in, until they are quite well developed and are not totally helpless in the presence of enemies.

The species living at the topmost level builds no such ponds – they might well be difficult to keep full of water at such a height. Their burrows, however, are very deep indeed and go down as much as a metre into the mud. They always contain some water at the bottom so these youngsters are also protected during the early stages of their lives.

Mudskippers, like fiddler crabs and oysters, are essentially marine creatures that have managed to adapt themselves to a life partly in the water and partly out of it. Some land animals, coming to the swamp from the other direction, have done much the same thing.

In southeast Asia, a small snake visits the swamps to hunt the mudskippers, chasing them across the flats and even pursuing them into their holes. It has become excellently adapted to living in water, having nostrils which it can close and a valve at the back of its throat that can be shut if it opens its mouth underwater to catch a fish. Another closely-related snake hunts not for fish but for crabs and has produced a venom that is particularly effective on crustaceans. A third snake has developed, most unusually, two movable tentacles on its nose which may help it to find its way through the muddy waters. In these swamps, too, lives a quite exceptional frog, the only one in the world that can tolerate salt water on its skin. It takes insects and small shrimps.

The most enterprising, inquisitive and omnivorous visitor of all is a monkey, the long-tailed macaque. It wades fearlessly on its hind legs, up to its waist in water if necessary, and has a particular fondness for crabs. Initially the crab is usually too quick for the monkey and scuttles down its hole, but the monkey will then sit just behind the hole and wait patiently until, at last, the crab cautiously peers out to see if it is safe to start feeding again – and the monkey grabs it. But it has to be careful how it does so, for the crab has pincers, and many a crab-hunt ends with the monkey squealing in fury and frantically flapping its hand in the air.

Twice every twenty-four hours the great arena of the mud is exposed to the air; and twice it is submerged. With speed and in silence, the water floods back. The tangle of roots disappears beneath a sheet of rippling water and the mangrove forest is transformed. To some of the mud dwellers – worms, crustaceans and molluscs – this brings relief. No longer are they vulnerable to aerial attack or in danger of being dried out. But not to all. Some of the crabs have become so adapted to breathing air that they cannot survive prolonged immersion. Each carefully builds a roof of mud over its hole that encloses a bubble of air with enough oxygen in it to sustain the crab until the water retreats again. Small mudskippers begin to clamber up the roots as though they were refugees from a flood. They are probably young ones that have not yet managed to claim territories for themselves on the mud below and so have no burrow in which to take refuge when large hungry fish come swimming into the swamp with the tide. Such youngsters may well be safer out of the water than in it.

Algae-browsing sea-snails also climb up the roots alongside the mudskippers. If they stayed on the mud, with no rock crevices to give them shelter, they too might be attacked by fish. They cannot, however, travel as fast as the mudskippers and have difficulty in keeping pace with the rise of the water. They therefore leave their feeding grounds on the mud long before the tide reaches them, displaying a remarkably accurate sense of time. Their internal clocks give them even more subtle alarms as well. Every month, the tides rise so high that the snails would not have enough time to travel all the way up from the mud to the height on the mangrove necessary to escape the tide. On these occasions, the snails do not descend to feed on the mud at all between tides, but crawl even higher up the mangrove roots to make sure that they keep out of harm's way.

Insects which had been feeding on the exposed flats have also been driven off by the water and settle in considerable numbers on the mangrove roots and under the leaves. But they are not completely out of danger. Among the marauding fish that swim into the mangroves with the flooding tide come archer fish, cruising close to the surface. They are quite large creatures, some 20 centimetres in length, and have big eyes and an upward jut to their mouths. Their vision is so acute that they are able to spot through the rippling, refracting surface of the water the insects perched above. Having sighted its prey, the archer presses its tongue against a long groove that runs along the roof of its mouth and jerks its gill covers together so that a stream of water shoots out like the jet from a water pistol. The fish may need two or three shots before it gets the range right, but it is persistent and in a high proportion of cases the insect is eventually knocked down into the water, and immediately gobbled up. The insects settled higher up the mangrove trees also have their attackers. The ghost crabs come up into the trees and turn over the leaves, picking off the sheltering flies with a snip of their pincers.

For several hours, the refugees on the roots are besieged. Then the ripples on the water vanish and for a few minutes everything lies still. The tide turns. The ripples reappear but trailing from the opposite side of the roots, and the swamp starts once more to drain. As the water flows away, it leaves behind yet another meal of edible fragments to be harvested by the crabs and the mudskippers, and yet another layer of sticky mud that extends the territory of the mangroves fractionally further into the sea.

Although the land is advancing in estuaries, elsewhere on the coasts it is in retreat. Where the shore is unprotected by deposits of sediments, and especially where it dips steeply away from the land, waves roll in and beat against the rocks. During a storm, they crash inshore with such force that they pick up boulders and sand and hurl them against the cliffs. The bombardment unfailingly exploits the lines of weakness in the cliffs – faults running up the face, layers of rock just marginally softer than the rest – eroding them more swiftly and deeply to create clefts and caves. As the land is cut back, segments become isolated as stacks and towers. The biggest boulders strike the cliffs most destructively at their base, undercutting them. Eventually, a whole face collapses. For a period, the mass of tumbled rock will protect the cliff base, but the sea picks up the debris, rolling the bigger boulders to and fro, crushing the smaller ones, grinding them into smaller and smaller pieces. In time, the fragments become so small that they are gathered up by the currents running along the coasts and carried away. Once more the base of the cliff is exposed to attack and once more the sea begins to grind its way inland.

Animals not only live in this dangerous demolition zone, but actually contribute to the destruction. The piddock, a bivalve mollusc like an elongated cockle, lives on softer rocks such as limestone, chalk and sandstone. The two valves of its shell are not connected by a horny hinge, as a cockle's are, but by a stout ball-and-socket joint. The animal puts out a fleshy foot from one end of the shell, grasps the rock and then pulls the sharp-toothed edge of the two valves on to the rock face. Then, by alternately scraping one valve after the other against the stone, rocking from side to side on its joint, it slowly gouges out a tunnel. Eventually it may excavate a shaft as much as 30 centimetres long and there it lies at the innermost end, with two conjoined siphons extending from the back of its shell down the tunnel to the entrance through which it sucks in and expels water, safe from the blows of rolling rocks – until, that is, the boulder it inhabits becomes riddled with so many holes that it splits. Then the piddock must start a new excavation elsewhere before it is crushed.

The date shell also bores into limestone, but instead of doing so mechanically it dissolves the rock with acid. Its shell, like that of any other mollusc, is made of the same substance as limestone, calcium carbonate, and would be vulnerable to its own acid were it not protected by a brown horny veneer which gives it its date-like appearance.

Cliffs and stacks, Port Campbell, Australia

The further up from the low-water mark a marine organism lives, the greater the stresses it has to withstand – the longer it has to remain out of water between tides, the more likely it is to be over-heated by the sun, and the greater the amount of unwelcome rainwater that may flow over it. This graduation of hazards produces a series of distinct zones, each dominated by the group of organisms that best deals with the particular combination of difficulties, so that a rocky shore is banded in a most striking way.

Rocks, unlike mud, offer a solid base for plants to establish themselves and most rocky shores are clothed with growths that are sometimes dismissively labelled seaweeds, but which are more properly called marine algae. It may seem odd, at first sight, that there are no plants in the sea comparable in complexity with the flowering plants of the land. But most of a land plant's tissues are devoted to dealing with difficulties that do not exist in the sea. A land plant must actively collect the water essential for life and distribute it to all parts of its body. It has to raise its crown high enough to avoid being overshadowed and robbed of its share of sunshine by its rivals. It must have mechanisms to bring male and female reproductive cells together and others to ensure that its fertilised eggs are distributed to new sites. So land plants have had to develop roots, stems and trunks, leaves, flowers and seeds. But in the sea, all these problems are solved by water. Water provides the algae with both the support and the liquid they require; it transports the sex cells when they are released; and it distributes the fertilised eggs. Since the algae have no vessels full of sap, the saltiness of the water causes no problems over the retention of their internal fluids. Marine algae, of course, just like any other plant except fungi, do need light and that does not penetrate very deeply into water. So for the most part, seaweeds either float or, if they attach themselves to the bottom, live where the sea is relatively shallow.

Just below low-water mark grow kelps and oar-weed – strap-like plants that, in some parts, form dense masses and grow many metres long, trailing their length in the light near the surface. The claw-like holdfast with which they grasp the rocks has no absorbent function as the roots of land plants have, and serves only as an anchor. These plants can tolerate a certain amount of exposure to the air at exceptionally low tides, but they do not flourish much above that level. Higher up the shore, their place is taken by the wracks, smaller plants with gas-filled bladders in their fronds that raise them when the tide comes in and keep them close to the surface and to the light. Above them again grow other kinds of wracks. The water here is never more than a few feet deep and these wracks are quite short and have no need of bladders to lift them. All these inter-tidal algae have surfaces that are slimy with mucus which retains moisture for a long time, and helps to prevent them being completely dried out. The species that lives at the highest level is able to survive exposure to the air for 80 per cent of the time. Although there are many other species of algae that grow on shores, it is the kelps that dominate most and give each zone its obvious character.

Some of the shore animals, in the same way, live in zones. At the topmost level, beyond the highest limit of the most drought-resistant wrack, beyond even the highest reach of the tides, where the only sea water to arrive comes as spray, live tiny acorn

Zones on the shore, northwest coast of the United States

barnacles. Clamped to the rocks, their minuscule shells pressed tightly together, they conserve very effectively indeed within their shells the little amount of moisture they need. Their food requirements are so small that they manage, almost unbelievably, to collect from the spray enough particles to sustain themselves.

Lower down, mussels may form a dense dark-blue band across the rocks. They cannot withstand such a long exposure to the air as the barnacles can, and this inability determines their topmost frontier. Their lower limit is fixed by the starfish that prey on them. The starfish's feeding technique is simple, slow, but devastating. It crawls over a mussel, enveloping it with its arms which are lined on their underside with rows of suckers called tube feet. Slowly it pulls the valves of the mussel apart. Then it pushes its bag-like stomach out of its mouth at the centre of its body, so that the stomach lining presses against the soft parts of the mussel and dissolves and absorbs them. These starfish are abundant on the sea floor below the reach of the lowest tides, where they eat molluscs of all kinds. Few mussels, in consequence, can survive there. But although the starfish can live for some time out of water, they cannot feed. So only a foot or so above low tide, conditions tilt a little in favour of the mussels, and several feet above low-water mark they are able to dominate the shore.

Mussels attach themselves to the rocks with bundles of sticky threads, but their grip is not very secure and in parts of the shore where the waves beat particularly heavily, they cannot maintain their hold. Their place may be taken by goose-necked barnacles, bean-sized creatures with bodies enclosed between two limy plates, and a long wrinkled stalk, as thick as a finger, which fastens them very firmly indeed to the rocks.

Many other organisms live in this inter-tidal zone alongside the barnacles and mussels, though in a less dominant way. Acorn barnacles, larger than those that live in the spray zone, encrust the mussel shells. Sea-slugs, molluscs without shells, feed on the barnacles. In hollows among the rocks, where water remains even when the tide is out, ranks of multicoloured anemones wave their tentacles. Globular sea urchins, spined like pin-cushions, move slowly over the boulders, gnawing at encrusting algae with teeth that project from the mouth in the centre of their undersides like the jaws that grasp a bit in a drill.

Although these zones, each with its own specific community of animals and plants, appear so distinct and with such sharp and definite boundaries, they are by no means permanent and unchanging. The organisms within them are always ready to take advantage of the slightest chance of extending their territories. A particularly severe storm may dislodge one or two of the mussels, creating a hole in the otherwise continuous carpet. Subsequent waves may then rip away great sheets of the shells. The tiny floating larvae of both mussels and barnacles, which are always present in the water, now have their chance to settle, and the goose-necked barnacles may well succeed in establishing an outpost in what was hitherto mussel territory.

On the northwest coast of America, one of the seaweeds has evolved a method of actively invading the mussel beds. It has a rubbery stem, half a metre high, which ends in a crown of dangling slippery fronds, making it look like a small palm tree. This

Kelp holdfast

unique crown is the device that enables it to attack the mussels. In the spring, a young plant may, exceptionally, manage to get a grip with its holdfast on the shell of a mussel. During the summer, when it is exposed at low tide, the sea-palm produces spores which trickle down its fronds and drip directly on to the surrounding mussels and lodge between them. When autumn storms come, a wave that in other circumstances might have caused the mussels little trouble may catch beneath the crown of the palm and rip it up. Because the plant's grip on the mussel is so much firmer than the mussel's on the rock, the plant carries the mussel with it. The small sea-palms in the mussel bed nearby are now able to spread quickly and establish the next generation of sea-palms on the newly exposed rock.

None of the individual organisms on rocky shores can have a very long life-expectancy, for in the end the incessant activity of waves reduces rocks to fragments. Coastal currents pick up these grains and sweep them away from the sites of their origin, ceaselessly sorting them into banks of uniform size, carrying them down the coast and, when the current slackens, dumping them in the lee of headlands and spreading them over the floors of bays.

Such sand-covered shores support fewer animals than any other part of the land's margins. Here, every wave of every tide stirs the surface of the sand to a depth of at least several centimetres so that no seaweed can establish itself. There cannot therefore be communities of plant-eating animals here. Nor are there rivers bringing down twice-daily deposits of food. Even those edible particles brought in by the wash of waves do not provide enough sustenance for many large creatures, for the beds of sand act in just the same way as filter beds on a sewage works. The surge of oxygen-laden water through the sand allows bacteria to flourish down to some depth. They rapidly break down and consume 95 per cent of all the organic matter brought in by the sea water. So no worms are able to live by eating sand as so many can do in estuaries by eating mud. The sandy shore inhabitants that seek food from the water must claim it before the sand-living bacteria do.

The sandmason worm does so by cementing together sand grains and fragments of shell to construct a tube that projects several centimetres above the surface of the sand. Its top end is fringed with tassels. These trap waterborne particles which the worm then picks off with its tentacles. Razor shells bury themselves in the sand for safety, but poke two tubes into the clear water above, through which they suck a current into a filter between their shells. The masked crab lives in a similar way. Lacking a mollusc's fleshy siphon, it improvises a suction tube by clasping its two antennae together. Several species of sea urchins have also become burrowers. Their spines are much shorter than those of their rocky-shore relatives. They use them for burrowing, twirling them on their ball-and-socket joints so that a digging urchin looks like some extraordinary miniature threshing machine. Once established below the sand, the animal covers the grains around it with mucus so as to create a small firm-walled chamber. Sea urchins, like starfish, possess tube feet. In these burrowing species, two tube feet are greatly elongated and snake upwards through shafts in the sand. Beating hairs that cover them waft water through the shafts so that the buried urchin can collect dissolved oxygen and food particles down one and discharge its waste up the other. Living a buried life, these urchins are seldom seen alive, but when they die their beautiful bleached skeletons are often washed up on a beach. The deeper digging species are heart-shaped; the ones that live near the surface are circular and flat, and are often called sand dollars.

The most plentiful food on the beach lies, inconveniently for most marine creatures, around high-water mark – where the waves cast up and abandon great quantities of organic debris – wrack and kelp ripped from the rockier parts of the coast, jellyfish blown inshore, bodies of dead fish, egg capsules of molluscs: the crop varies from tide to tide and season to season. Sand-hoppers, relatives of shrimps, are able to obtain all the moisture they need from damp sand and, for most of the day, hide themselves in it beneath the moist piles of cast-up seaweed. When night falls and the air cools, they swarm out, as many as 25,000 of them in one square metre, and begin demolishing the rotting weed and flesh. But they are exceptional. These rich pickings are beyond the reach of most of the sea creatures that live on the beach.

On the southern shores of Africa, however, there lives one mollusc, the plough snail, that has perfected the most ingenious way of visiting this source of food with the minimum of effort and risk. It lies buried in the sand just around low-water mark. As the tide washes over its hiding place and advances up the shore, the plough snail emerges from the sand and sucks water into its foot. This expands into a large structure shaped like a ploughshare. It serves the animal not so much as a plough but as a surf-board. The waves catch beneath it and sweep the snail up the beach, depositing it at the same level as they drop much of their flotsam. The snail is extremely sensitive to the taste of decomposition in the water. As soon as it detects it, it retracts its surf-board and crawls across the wave-washed sand to the source of the taste. A stranded jelly-fish will attract dozens within a few minutes. They begin feeding immediately, before the tide reaches its highest point and while their prey is still awash. To travel to the uppermost limit of the waves would be dangerous, for then, if the snails fed for any length of time, they might miss their lift back down the beach and be stranded. As the waves creep higher, they abandon their feeding and dig into the sand. Only when the waters start to retreat will they emerge, reflate their fleshy surf-board, and be carried back to deeper waters to wait beneath the sand for the next tide.

Very few marine creatures can venture above the limit of the highest tide and survive. Turtles, however, are compelled to do so by their ancestry. They are descended from land-living, air-breathing tortoises, and have, over the millennia, become superb swimmers, able to dive beneath the surface for long periods without drawing breath, propelling themselves at great speed with legs that have been modified into long broad flippers. But their eggs, like all reptilian eggs, can only develop and hatch in air. Their developing embryos need gaseous oxygen to breathe, and without it they will suffocate and die. So every year, the adult female turtles, having mated at sea, must leave the safety of the open ocean and visit dry land.

Plough snails feeding on jellyfish, South Africa

Ridleys, one of the smallest of the sea-going turtles, averaging about 60 centimetres in length, breed in vast congregations that must surely constitute one of the most astounding sights in the animal world. On one or two remote beaches in Mexico and Costa Rica, on a few nights between August and November which scientists so far have been unable to predict, hundreds of thousands of turtles emerge from the water and advance up the beach. They have the lungs and the watertight skin bequeathed them by their land-living ancestors so they are in no danger of either suffocation or desiccation, but their flippers are ill-suited to movement out of water. But nothing stops them. They struggle upwards until at last they reach the head of the beach just below the line of permanent vegetation. There they begin digging their nest holes. So tightly packed are they that they clamber over one another in their efforts to find a suitable site. As they dig, sweeping their flippers back and forth, they scatter sand over one another and slap each others' shells. When at last each turtle completes her hole, she lays about a hundred eggs in it, carefully covers it over, and returns to the sea. The laying continues for three or four nights in succession during which time over 100,000 Ridleys may visit a single beach. The eggs take forty-eight days to hatch but often before they can do so a second army of turtles arrives. Again the beach is covered with crawling reptiles. As the newcomers dig, many inadvertently excavate and destroy the eggs laid by their predecessors and the beach becomes strewn with parchment-like egg-coverings and rotting embryos. Less than one in 500 of the eggs laid on the beach survives long enough to produce a hatchling.

The factors that govern this mass breeding are still not properly understood. It may be that all the Ridleys come to these very few beaches simply because ocean currents happen to sweep them passively in that direction. It could also be advantageous for them to nest thousands at a time, since if they spaced out their visits more evenly throughout the year, the beaches would attract large permanent populations of predators such as crabs, snakes, iguanas and vultures. As it is, there is little to eat there for most of the year, so very few such creatures are there when the turtles arrive. If this is the reason for the habit, it seems to have been effective, for Ridleys in both the Pacific and the Atlantic are among the commonest of all turtles while many other species are now diminishing in numbers and some are in danger of extinction.

The biggest of them all is the leatherback turtle which can grow to over 2 metres in length and weigh 600 kilograms. It differs from all other turtles in that it has no horny plated shell but a ridged carapace of leathery, almost rubbery skin. It is a solitary ocean-going creature. Individuals may turn up anywhere in tropical seas and have been caught as far south as Argentina and as far north as Norway. Until only twenty-five years ago, no one knew where the species' main nesting-beaches were. Then two sites were discovered, one on the east coast of Malaysia and another in Surinam in South America. On each, the leatherbacks nest, a few dozen at a time throughout a three-month season.

The females usually come at night when the moon has risen and the tide is high. A dark hump appears in the breakers, glistening in the moonlight. With strokes of her

Overleaf: *Mass breeding of Ridley turtles, Costa Rica*

immense flippers, she heaves herself up the wet sand. Every few minutes she stops and rests. It may take her half an hour or more to climb to the level she seeks, for the nest must be above the reach of the waves yet the sand must be sufficiently moist to remain firm and not cave in as she digs. She may make several trial excavations before she finally discovers a place that suits her. Then she determinedly starts clearing a wide pit with her front flippers, sweeping showers of sand behind her. After a few minutes work, it is deep enough, and with the most delicate movements of her broad back flippers, she scoops out a narrow shaft in the bottom.

She is almost entirely deaf to airborne sounds and your talking will not have disturbed her. Had you shone a torch on her as she climbed up the beach, however, she might well have turned around and gone back to the sea without laying. But now even bright lights will not stop her getting rid of her eggs. She sheds them swiftly, in groups, her back flippers clasped on either side of her ovipositor, guiding the eggs downwards. As she lays, she sighs heavily and groans. Mucus trickles from her large lustrous eyes. In less than half an hour, all her eggs are laid and she carefully fills in the pit, pressing the sand down with her hind flippers. She seldom returns directly to the sea but often moves over to other places on the beach where she digs in a desultory way, as though to confuse her trail. Certainly, by the time she heads back to the waves, the surface of the beach has been so churned up that it is almost impossible to guess just where her eggs lie.

But watching human beings need not guess. In Malaysia and in Surinam, the people patrol the beaches all night and every night during the season and collect the eggs, usually even before the females have filled in the shaft. A few eggs are now being bought by local government agencies and hatched in artificial incubators, but nearly all the rest are sold in local markets and eaten.

Perhaps we have not yet discovered all the leatherbacks' breeding grounds. Maybe some of the turtles, as they wander through the seas, have come across remote deserted islands, far from the haunts of man, where they can breed unmolested. They are not the only creatures to make such voyages. Shore-living organisms are unable to move from shallow water during their adult lives; but at an earlier stage they too travel widely, floating as seeds and larvae, eggs and fry. For them all, an island may be not just another densely populated, highly competitive home like the coasts from which they have come, but a sanctuary where they can have the freedom to develop into forms that are quite new.

TEN

WORLDS APART

If you wanted to find the loneliest of islands, far from the paths of shipping and cut off from the rest of the world, you might well choose Aldabra. It lies in the Indian Ocean, 400 kilometres east of Africa and about the same distance north of Madagascar. To sail to it, your navigation must be good, for Aldabra is only 30 kilometres long and, even at its highest point, only 25 metres above sea level. So you will not see it until you are quite close. Indeed, your best chance of locating it by eye is to look, not for the land itself, but for the pale green light reflected from its vegetation and the green waters of its shallow lagoon on to the underside of clouds that hang above it. If you have come from Africa and miss it, you may sail on for several days more until you sight one of the southern Seychelles; and if you miss them and continue on the same course you will see no more land until you reach Australia, 6000 kilometres away.

Aldabra is an atoll, a coral cap on a submarine volcano that rises steeply from the sea floor 4000 metres below. In form it is a huge lagoon, encircled by a ring of islets separated from one another by narrow channels. The surface of these islets is coral rock, eroded by the chemical action of rain into a honeycomb of jagged blades and deep clefts. Enclosed within this limestone are layers of rubbly sandy soil, and from these we know that the atoll has risen and fallen above and below the surface of the water several times as the level of the sea or the contours of the ocean floor have risen and fallen. It last emerged about 50,000 years ago. As the reefs slowly rose, the waves broke over them less and less frequently until eventually the limestone drained and a new island was established. At that moment, of course, it can have had no land-living inhabitants whatever, but as the millennia passed, animals and plants of many kinds began to arrive by sea and air, so that today Aldabra is the home of a large and varied community.

Sea birds, not surprisingly, go there in vast numbers. Reaching even such a remote island as Aldabra presents no problems to such accomplished travellers, and at certain times of the year the sky above the atoll is filled with wheeling flocks of red-footed boobies and frigate birds.

Both birds depend on the surrounding seas for their food. The boobies, a kind of gannet, range for several hundred kilometres in all directions. When they find shoals of

Overleaf: *Frigate birds, Aldabra*

fish or squid, they rocket down into the water, plunging several metres below the surface to seize their prey. The frigates, huge black birds with a 2-metre wing-span and a deeply forked tail, have a slightly different fishing technique. They skim at high speed close to the surface of the sea and, with a nod of the head, deftly pluck a squid or a flying fish from the water with their long hooked bills. But they also soar around the island waiting for the returning boobies. When one arrives with a crop full of fish, the frigates harry it so persistently that often the booby is forced to disgorge its meal, whereupon one of the frigates will swoop and catch the fish before it hits the water.

Both boobies and frigates spend most of the year in the air, seldom settling on the water. They come to Aldabra in order to nest. Islands free of cats, rats and any other creature that might suck eggs or eat chicks are not numerous and Aldabra serves as the breeding headquarters for all the frigates of the Indian Ocean. They come here from as far away as the coast of India over 3000 kilometres away and they nest in the low mangroves at the east end of the island. The males are the first to settle. They sit among the branches and inflate huge scarlet throat pouches that balloon beneath their bills. These are invitations to the females flying above to join them and nest.

The boobies, notwithstanding their treatment by the piratical frigates, nest alongside. As there are no predators on the island, the birds have no need to conceal their nests or to place them in inaccessible positions. So the mangroves become laden with nests often constructed so close together that one building bird can steal sticks from its neighbour's without even moving from its own.

As well as mangroves, there are many other plants here. Coconut palms fringe the beaches. Low thorn bushes manage to take root in crevices in the coral limestone. And in places where sand has blown inland, there is a low green turf. How did all these plants reach here? Some seeds certainly arrived by air, stuck to beaks, feet or feathers of birds. Some may even have travelled within a bird's stomach and have been squirted ashore with its droppings. Smaller seeds may have been driven across from the mainland in a storm, supported by their own tiny parachutes of down. Many of the rest will have come by sea. Walk along the beach at high-water mark and you can pick up, within a few yards, half a dozen different kinds of seeds that have been dumped there by the waves. Some may be dead, but many will still be viable and a few may already be sprouting roots and leaves.

Coconuts are common among them and indeed the coconut palm is one of the most successful practitioners of this way of travelling around the world. It grows naturally along beaches of tropical islands in the narrow strip just above high-water mark, where it is not shaded by land trees or choked by undergrowth. It leans out over the beach so that its nuts, when they fall, tumble within reach of the waves which then wash them down the beach and out to sea. There they float, supported by the thick husk of coarse fibre which surrounds their hard-shelled, flesh-filled seed. They can remain alive at sea for as long as four months, during which time they may travel hundreds of kilometres before being cast up on a new beach, a still uncolonised Aldabra. So successful has the coconut been in distributing itself around the world – and so assiduous has man been in

Coconut sprouting on a tropical beach

planting such a valuable source of food and drink – that now it is probably impossible to discover exactly where the species originated.

The flotsam at high-water mark also contains pieces of driftwood, tangles of roots and all kinds of vegetable trash. Though these fragments are themselves dead, they may, nonetheless, have brought life with them in the form of animal passengers. Many of Aldabra's snails, millipedes, spiders and other small invertebrates must have arrived in such a way. Even bigger animals may have made the voyage in that fashion. Reptiles are particularly hardy sailors, capable of surviving long voyages on such rafts. Amphibians, on the other hand, lacking the reptile's water-tight skin, cannot survive soaking by sea water. So Aldabra, like nearly all oceanic islands, has many lizards skittering around its trees and warming themselves on the rocks, but no frogs calling from its brackish swamps.

Animals and plants, having arrived on an island, do not all remain the same, generation after generation. As time passes, many species begin slowly to change. The process is the same as that which produces new species among populations of fish isolated in lakes. Tiny alterations occur in the genetic structures during the complexities of reproduction and these produce anatomical differences in the offspring. In small in-breeding communities, such mutations are not diluted as they might be in a bigger population on the mainland, and so they are more likely to find a permanent place in future generations. So evolutionary change proceeds particularly swiftly on islands as it does in great lakes.

Sometimes, the changes seem to be quite trivial. Aldabra's most handsome flowering plant, a species of *Lomatophyllum*, produces, from the centre of its rosette of spiny succulent leaves, tall spikes of orange blooms which are slightly different in colour from the *Lomatophyllum* that grows on the Seychelles, a few hundred kilometres away. Similarly, Aldabra's single resident bird of prey, a beautiful little kestrel, has a slightly redder underside than its cousins in Madagascar, a difference which is significant and constant enough to justify classifying the bird as a separate sub-species.

Other changes among the island's residents, however, are more substantial. Scurrying through the bush, so incurably inquisitive that you can summon it just by tapping a rock with a pebble, is a little rail. Rails are small, long-legged birds related to moorhens and coots, and Aldabra's rail is not unlike its relative that lives on the mainland of Africa. It has the same build and the same sort of habits. But there is a major difference between the two. The Aldabra rail cannot fly.

In Africa, rails rely on their ability to take to the air to escape the multitude of hunting animals that would otherwise attack them. Here on Aldabra, there are no such dangers, so the loss of flight is no handicap. On the contrary, it brings positive advantages. Flight makes great demands. The muscles and bones necessary to beat wings effectively make up about 20 per cent of the weight of a typical bird, and developing it all requires a great deal of nourishment. Each time a bird lifts itself into the air it burns up a great deal of energy, so it is little wonder that birds seldom fly unless they have a real need to do so. The rails on Aldabra have none. In fact, there might even be danger in doing so, for strong winds blow across the island for much of the time, and the birds

Flightless rail, Aldabra

might easily be carried out so far to sea that they would have real difficulty in returning. So their wings are small and poorly muscled and they hardly ever even flap them.

The Aldabran rail is not the only member of its family to have reacted to isolation in this way. On Tristan da Cunha, Ascension and Gough Island and on several islands in the Pacific, there are rails that either cannot fly or are only just able to flutter into the air. In New Caledonia, in the western Pacific, a relative of the cranes has become flightless. This is the kagu, a strikingly beautiful bird with a resplendent head plume and a spectacular courtship dance during which it proudly displays its ineffective wings to its mate. In the Galapagos, the same evolutionary path has been taken by cormorants whose stunted wings have feathers so tattered that they are quite incapable of lifting the birds into the air, no matter how hard they flap them.

But perhaps the most famous of all flightless island-living birds is one which once lived on one of Aldabra's distant neighbours in the Indian Ocean, Mauritius. This was a pigeon-like bird that took to foraging on the ground for its food and grew to the size of a large turkey. Its body feathers became soft and downy and its wings no more than vestiges. Its tail, once an aerodynamically serviceable fan, was reduced to a decorative curling tuft on its rump. The Portuguese called it *doudo*, simpleton, because it was so trusting that it could easily be hit over the head and killed. Sailors of all nationalities slaughtered the dodos in great numbers for food. Pigs, introduced on to the island, ate their eggs. The last of the dodos had been killed by the end of the seventeenth century, less than 200 years after travellers had first described them.

Two other flightless pigeons lived on nearby islands, one on Réunion and one on Rodriguez. Apparently of different species, European seamen called them both solitaires as they were found singly in the forests. They were about the same size as the dodo but with longer necks and, from all accounts, a much more stately gait than the waddling dodo. They, too, were both exterminated by the end of the eighteenth century.

Feeding alongside the dodos and solitaires in Mauritius, Réunion and Rodriguez were also huge tortoises. They grew to over a metre long and weighed up to 200 kilos. Others lived on the Comoro Islands and Madagascar. These were even more valuable to sailors than the dodo, for they would remain alive in the holds of ships for weeks on end, so they could provide fresh meat, even in the Tropics, many days after the ship had last left port. So the giant tortoises went the same way as the dodo and the solitaire. By the end of the nineteenth century, all the giant tortoises of the Indian Ocean had been exterminated – except for those on Aldabra. They were so isolated and so far from the main shipping routes that even the prospect of conveniently-packaged fresh meat did not tempt many captains to go so far out of their way. Today, there are still some 150,000 giant tortoises on the island.

There seems to be no doubt that they, like their extinct relatives on islands elsewhere, are descended from normal-sized tortoises living on the African mainland. It may be that some of these, many thousands of years ago, made the trip across to Madagascar riding on clumps of vegetation. It is also possible that, once giant forms developed, they spread to other islands supported by no more than the buoyancy of their own bodies.

Kagu, New Caledonia

Giant tortoise eating from another's corpse, Aldabra

Such voyages may well have started when a tortoise, grazing among the mangroves by the edge of the sea, was accidentally caught by the tide and swept out. Giant tortoises have certainly been found floating among the waves many miles from land and they can probably survive many days at sea. An ocean current flows from Madagascar towards Aldabra and, helped by that, a giant tortoise could make the passage in about ten days.

It is by no means certain why tortoises, when isolated on islands, should become giants. Maybe a large animal, with big reserves of fat, is better able to survive a bad season than a smaller one. There may be an even simpler reason. With no predators around to attack them and no animals competing with them for the grazing, naturally long-lived creatures like tortoises may just go on growing.

As well as increasing their size, the island tortoises have changed in other ways. Pasturage is not abundant on many of these islands and it is particularly scarce on Aldabra. Tortoises have consequently broadened their diet to include almost anything that is remotely edible, as you will soon discover if you camp on the island. The animals not only sit expectantly around you at mealtimes but will slowly and ponderously demolish your tent in the search for something to eat, as well as sampling any bits of your clothing that you may have left lying around. More grimly, they have also become cannibals. When one of their number dies, these normally vegetarian creatures can be seen champing their way through the decomposing entrails of the corpse.

The relative proportions of their bodies have also changed. Their huge shells are not as thick or as strong as those of their African relatives, nor are the internal struts of bone that support the carapace as robust. Indeed, their shells are easily dented if they are roughly handled. Nor does the carapace provide such an effective refuge as that of the mainland tortoise. The opening at the front has become wider and the animal's body bulges farther from it. This gives the tortoise much greater freedom while browsing, but it also means that the animal is unable to withdraw its limbs and neck completely into the shell. Were it to be transported back to Africa, hyenas or jackals might well be able to fasten their teeth into the tortoise's neck and kill it.

Tortoises on isolated islands elsewhere in the world have changed in a very similar way. There are some in the Galapagos that are equally big. Their closest relations, however, are not the giants in the Indian Ocean, but tortoises a fraction of their size that live in South America.

This tendency of island-living reptiles to grow huge is not limited to tortoises. A lizard in Indonesia has evolved in much the same way. Its main home is Komodo, a small island, a mere 30 kilometres long, in the centre of the Indonesian chain. This lizard, popularly called the Komodo dragon, is in sober fact a species of monitor, closely related to the goannas of Australia and the water monitor which is common in many tropical countries from Africa to Malaysia. But it does grow to a length of 3 metres, which is considerably longer than any other lizard. It is also more massive, for whereas some two-thirds of the length of other monitors is taken up by their tail, in the dragon this proportion is only about half. So even a half-grown dragon is much more bulky and formidable than any other monitor of the same length.

Overleaf: *Komodo dragon*

The dragon is a meat-eater. When it is young, less than a metre long, it preys on insects and small lizards, clambering around in trees to collect them. Half-grown, it hunts almost entirely on the ground and catches rats, mice and birds. Fully adult, it lives predominantly on the flesh of pigs and deer that occur naturally on the island as well as that of goats that have been introduced there by man. Much of this meat is certainly taken as carrion, but the dragon is also an active hunter. It will follow pregnant goats and seize a newly-born kid as soon as it drops to the ground. It also waits in ambush for adults, lurking in thickets beside the game trails habitually used by goats, pigs and deer. As one passes, the dragon seizes it by the leg with its jaws and after a struggle, throws it to the ground. Before the victim can recover, the dragon rips open its belly and it quickly dies.

Stories are told on the island of dragons attacking human beings and it seems certain that, in the past, one or two people have stumbled across an animal, been badly bitten, and later died of their wounds. There is little doubt, too, that if a man were to collapse in the heat, a dragon would treat his body as it would any other animal carcass. But it seems unlikely that they regularly regard human beings as prey. Sitting in the bush, watching them, you do not feel that you are being hunted. A dragon is likely to watch you just as long as you watch it, lying as motionless as a statue, except for a blink, an occasional sigh or a flick of its long, yellow forked tongue with which it savours the scent-laden air. Even when it heaves itself to its feet and advances towards you in a purposeful way, it seems happy to plod straight past you.

But when you see dragons gathered around a carcass, their potential ferocity and strength become more apparent. A large one is fully able to pick up a goat's carcass with its jaws and drag it bodily away. If two large ones are feeding on it, they each fasten their jaws in it and rip it apart with backward jerks of their head and shoulders. If younger ones are rash enough to dispute the food with their elders, they are driven away with a lunging rush. Nor are these sham attacks. Analyses of droppings show that the adults regularly eat the smaller ones. The dragon is also a cannibal.

The diet of these lizards may be one reason why they have grown into giants. There are no other large meat-eaters here, no other hunters preying on the deer and the pigs grazing in the bush. It seems likely that the dragon's ancestors fed on carrion, insects and small mammals, just as other smaller monitors do elsewhere and young dragons do in Komodo. But, eventually, some became large enough and strong enough to tackle the living herbivores and so were able to tap a rich and otherwise unexploited source of meat. Finally, this characteristic became common to the whole species and the Komodo dragon evolved into the largest lizard in the world.

Today, the dragons live not only on Komodo but on the neighbouring islands of Padar and Rintja and the western end of the very much bigger island of Flores. They are capable swimmers and are regularly seen crossing narrow straits to hunt on Komodo's offshore islets, but it is not certain whether they have spread among the islands in this way. It might also be that the land, in this volcanic region, has sunk within recent geological time and the large island that was the lizards' original home has now become divided by the sea into the several smaller islands that we see today.

Most island species do remain within the islands on which they evolved, so they often become the subject of tall tales related by the few travellers who have visited their lonely homes. The Komodo dragon owes its popular name to such romantic exaggerations and when the world first began to hear about its existence, at the beginning of this century, stories circulated maintaining that the monster grew to a length of 7 metres, over twice its actual size. Five hundred years ago, an island-living plant gave rise to even more marvellous stories.

Then as now, immense nuts, like two huge coconuts joined side by side, were, very rarely, washed up on the shores of the Indian Ocean, usually in their huge boat-shaped husk. Arabs found them and treasured them. So did Indians and the people of southeast Asia. But no one knew what tree produced them. The nuts themselves could not provide the answer by germinating and sprouting, for all of them, without exception, were dead. The most widespread belief was that they were the fruit of a tree that grew beneath the surface of the sea. So the nut was called a coco-de-mer.

To the imaginative eye, these nuts, with their central cleft, resembled a female pelvis. At any rate, they were widely credited with aphrodisiac powers. Drinks made from their hard kernels were, it was said, irresistible love potions. The shells, too, were magical. Cups made from them would render even the most powerful poison completely harmless. So a coco-de-mer nut became worth a king's ransom and throughout the Orient, and even in the royal courts of Europe, they were elaborately carved and mounted in silver and gold.

It was not until near the end of the eighteenth century that the tree which produced the nuts was discovered. It grew in the Seychelles, on the islands of Praslin and Curieuse. The trees are as spectacular as their fruit. On Praslin, they grow in a dense grove. The huge trees, many of them several centuries old, have straight smooth trunks rising, clear of branches, for 30 metres. Their leaves are immense pleated fans, 6 metres across. Each tree carries only male or female flowers. The female trees stand taller than the males, their crowns hung round with clusters of gigantic nuts which take seven years to mature. The smaller male trees produce their flowers on a long chocolate-brown spike. And on almost every one of these sits an exquisite jewel-like lizard, a gecko of brilliant emerald green spangled with scales of delicate pink. It, too, is a unique creature. Its family, Phelsuma, originated in Madagascar but Praslin, like other Seychelles islands, has its own particular race with its own special colouring.

The coco-de-mer nut is the largest seed produced by any plant. Unlike a normal coconut, which is hollow when mature, the coco-de-mer is completely filled with hard flesh. This makes it so heavy that the nut will not float high in the water like a coconut. Indeed, salt water kills it. So this island plant is not only restricted to this small group of islands, but must have evolved there, or on a larger tract of land, much of which has now been submerged and of which these small islands are the only surviving relics.

The islands we have examined so far have been relatively small. Their populations have each evolved along a single line to produce a single new species. So there is only

one kind of giant lizard on Komodo, one giant tortoise on Aldabra and there was only one dodo on Mauritius. But if the island is large and contains within it a variety of different environments, or if there is a cluster of islets, each with its own special character, then a single invader may evolve into not just one new form, but a multitude.

The most famous examples of this phenomenon are those birds that Darwin observed to such effect, the finches of the Galapagos. Many thousands of years ago, a flock of finches is believed to have been blown by a freak storm away from the coast of South America, out into the Pacific. This must have happened many times, both before and since, but this particular flock, luckily, found refuge at last on these volcanic islands, nearly 1000 kilometres out from the mainland. The islands must have already been colonised by plants and insects, for there was sufficient food for the wandering finches, and here they have stayed. The Galapagos islands, however, are varied. Some are very dry, with little more than cactus growing on them. Others are comparatively well-watered and have grassy plains and patches of thick bush. Some are low. Others have volcanic peaks that rise to 1500 metres, with rain-drenched valleys where ferns and orchids grow. So a number of different environments were available to the finches. No other birds were exploiting them. No woodpeckers were excavating grubs from the tree bark, no warblers collecting insects, no pigeons pecking at fruit. As time passed, different populations of finches became more and more skilful at extracting food from a particular habitat. They did so by modifying the instrument with which they collected it – their beak.

Today, on one single island, there are ten different species of finch. In size, body shape and feather colour, they are all much the same. Only in their beaks and their behaviour do they differ to a very marked degree. One which feeds largely by crushing buds and fruit, rather as a European bullfinch does, has a hefty deep beak. Another, which feeds by picking off small insects and larvae, has a slender beak which it uses with great delicacy and accuracy, like a pair of tweezers. A third has a medium-weight, sparrow-like beak and feeds on seeds. And one, as though impatient at the length of time it takes for evolution to modify physically any part of its anatomy, has instead modified its behaviour and taken to tool-using. It carefully cuts a spine from a cactus and then uses it to extract beetle grubs from their holes in rotten wood with all the skill of someone taking a winkle out of a shell with a pin. Altogether, there are now fourteen different species of finch in the Galapagos archipelago.

In the Hawaiian islands, the process has progressed even farther. They are even more remote from continental land than the Galapagos, being 3000 kilometres from the coast of California; they are even bigger, with a greater variety of environments; they are geologically older; and their animal colonists reached them much earlier.

Their most characteristic group of birds are the honey-creepers. They are sufficiently similar to make it reasonably certain that they are all descended from the same ancestral group, but they have now changed so much that it is extremely difficult to decide exactly what that ancestor was. It may have been a finch, or maybe a tanager. They vary, not only in beak shape like the Galapagos finches, but in colour as well. Some are

Young leaves of coco-de-mer, Praslin, Seychelles

scarlet, others green, yellow or black. As for their beaks, some are parrot-like and used for cracking seeds, others long and curving which enable their owners to probe deep into Hawaii's lovely flowers to sip nectar. One insect-feeding species had a beak with mandibles of different lengths, the upper curved and serving as a probe and the lower short and straight like a dagger with which it chipped wood. When man first came to these islands there were at least twenty-two different species of honey-creeper. Sadly, almost half of them are now extinct.

Apart from the honey-creepers, representatives of only five other families of birds are found in the Hawaiian archipelago. Compare that with the fifty-odd families that live in Britain. The explanation is not only that Hawaii is so extremely isolated, but also that the honey-creepers were the first birds to colonise the islands, and when, in more recent times, other potential bird settlers arrived, as they must assuredly have done, they stood little chance of establishing themselves. Most ecological niches were already occupied by different kinds of honey-creepers.

Hawaii and the Galapagos are both of volcanic origin. When they first arose from the sea, they offered completely vacant land to the first colonists who arrived there by sea and air. Aldabra did the same. But other islands have originated in a different way. They are parts of a continent that became isolated when the land sank beneath the sea, leaving only the tops of mountains exposed as islands, or when splinters were pulled away from the main body of the continent by the moving plates of the ocean floor. Such islands as these often carry with them into isolation a whole ark-load of involuntary passengers. They then become not only nurseries for new species but sanctuaries for ancient ones.

This process took place on a continental scale some 100 million years ago, when the great southern supercontinent began to split up into South America, Antarctica and Australasia. At this time, amphibians and reptiles were widespread and birds were well established. New Zealand drifted away very early in the break-up and took with it representatives of all these groups. Subsequently, marsupial mammals evolved in Australia, changing the whole balance of the animal community there. But they could not reach New Zealand, and primitive amphibians and reptiles survived there longer.

Three species of small primitive frogs can still be found by those who search carefully in the cool damp forest, and lizards, both skinks and geckos, are quite common. But one reptile is of exceptional interest – the tuatara. Outwardly, it looks like a rather heavily-built lizard. Its true character only becomes obvious when its skeleton is examined. Its skull bones show that this small sluggish creature, a mere foot or so long, is closely related not to modern lizards but to the dinosaurs. It is the most ancient of living reptiles and the fossilised bones of a virtually identical creature have been found in rocks 200 million years old.

The forests of New Zealand, themselves made up of trees of a very ancient kind, kauri pine, southern beech and tree ferns, also contain another animal survivor from early times, the kiwi. It is a bird the size of a chicken with powerful digging legs and a long beak with which it probes for worms. Its feathers are elongated to such a degree

The i'iwi honeycreeper, Hawaii

that they are almost hair-like, and its wings are so small that they are virtually hidden beneath its coat of feathers. It is the last survivor of a whole tribe of flightless birds, the moas, that once lived on the islands. From their bones we know that there were at least a dozen different species. Some were small, little bigger than the kiwi, and grazed on the forest floor. Others grew very tall indeed. The biggest stood 3.5 metres high and were the tallest birds that have ever existed. They were also vegetarian, as we know from the piles of worn gizzard stones that have been found within the ribs of their skeletons, and probably browsed on trees. It seems that, in the absence of any herbivorous mammals, these flightless birds took the places occupied elsewhere in the world by large rodents, deer and even giraffe.

Great flightless birds are found in many parts of the world – the ostrich in Africa, the rhea in South America, the emu in Australia, as well as the extinct elephant bird which lived in Madagascar and which, while not as tall as the biggest moa, was heavier than any of them. It may be that all these birds lost their ability to fly a very long time ago when the great southern supercontinent was still unfragmented. Each, at that time, was so large and powerful that it was able to maintain itself, even when ferocious predatory mammals appeared. If that was the case, then the ancestral moas must have been living alongside the tuatara and the primitive frogs when New Zealand separated from Australia.

There is an alternative explanation. Maybe the ancestors of the moas were still able to fly at the time of New Zealand's separation and only subsequently became ground-living and gigantic in isolation, just as the dodo and the solitaire did. However that may be, many other birds certainly reached New Zealand by air. Many came from Australia, helped on their journey by the trade winds that blow regularly and powerfully in an easterly direction. Even now, avocets, cormorants, ducks and other vagrant Australian birds regularly appear in New Zealand. Those that landed and settled many thousands of years ago have since evolved in their own special way, just as birds did in Aldabra, the Galapagos and Hawaii. Here in New Zealand, the processes have been in action for longer still and produced wrens, parrots and ducks that are markedly different from all their relatives elsewhere in the world.

Fifty species of New Zealand's land birds are unique. Of these, fourteen have become either poor flyers or totally flightless. It is no surprise to find among them a flightless rail. This is the weka, a partridge-sized bird that runs through the forest and feeds on insects, snails and lizards. Another member of the rail family, the takahe, is a kind of coot that has not only lost its ability to fly but has also grown very large. It is the size of a small turkey and has a massive scarlet bill and bright-blue body feathers. Even more remarkable, one of New Zealand's parrots has also become flightless. The kakapo, sometimes called the owl parrot, has mossy green plumage and an engagingly solemn expression. It comes out at night to nibble fern leaves, mosses and berries. Although it can, albeit reluctantly, flap its way a few feet into the air or glide down a hillside, it is primarily a walker and a climber and maintains long paths through the moorland vegetation, clipping them clear when necessary, and here and there excavating little

Tuatara, New Zealand

amphitheatres against a rock or beneath a tree where, during the breeding season, it performs its courtship rituals and sings its booming song.

Between them, the animals of New Zealand illustrate all the effects of isolation. A great number have evolved into unique forms. Many, whose ancestors could once fly, have become ground-dwellers like the kakapo. Some, like the moas and the takahe, have become giants of their kind. But New Zealand, alas, also demonstrates in a most vivid way one further characteristic of island creatures – their vulnerability. They easily succumb to invaders.

The most lethal intruder is man. New Zealand remained unknown and unvisited by human beings until about a thousand years ago. The first people to reach it were Polynesians. They were, and are, among the greatest seamen the world has seen. Long before Columbus crossed the Atlantic, the Polynesians were visiting the scattered archipelagos of the Pacific. They probably made their first colonisations by relatively short journeys, moving out from the mainland of Asia, from one archipelago to another, into the heart of the Pacific. Then, from their headquarters in the Marquesas they made, over the centuries, a series of enormous voyages, north to Hawaii, west to Tahiti, east to Easter Island and eventually – the longest voyage of all – 4000 kilometres southwest to New Zealand. These were not accidental journeys caused by a sudden storm blowing a craft off-course. They were carefully planned. The canoes they used were immense double-hulled craft capable of carrying hundreds of passengers. When they were launched on colonising voyages, they carried women as well as men, and had in their holds roots of food plants, domestic animals and all the other things the people needed to found a new self-sufficient community.

New Zealand must have held one great and welcome surprise for the Polynesians. None of the islands they had occupied until then had any large animals living on them. They had had to rely for meat on the pigs and chickens they had brought with them. But New Zealand had a large population of giant birds, the moas, and the Polynesian colonists, the Maoris, hunted them with vigour and success. They not only ate the flesh of the moas, they used the skins for clothing, the eggs as containers and the bones for the tips of weapons and tools and as jewellery. The middens outside ancient Maori villages have yielded vast numbers of moa carcasses. Without any doubt, this hunting must have greatly reduced the numbers of the moas. But the Maoris also began clearing the forest, which at that time covered the greater part of the islands. As it was cut down and burnt, the moas lost not only their browse but their hiding places. The Maoris also brought with them dogs and a Polynesian rat, the kiori. Both must have taken their toll of the birds, if not as adults, then as chicks or eggs. Within a few centuries of the Maoris' arrival, all the moa family, except the kiwi, were extinct. Nor were they the only birds to be affected. Of the 300 species that are believed to have existed on the islands before man arrived, forty-five disappeared.

Then, 200 years ago, Europeans arrived. They caused more havoc. They brought, in their ships, another species of rat. They cut down further vast areas of forest, turned it into grassland and grazed immense flocks of sheep on it. Apparently, they found little

Kakapo, New Zealand

pleasure in the outlandish island creatures that they discovered in residence and so they introduced more familiar long-loved animals to remind them of their original home. Societies were formed with this specific aim. From Britain they brought mallards and skylarks, blackbirds and rooks, chaffinches, goldfinches and starlings; from Australia, black swans, kookaburra and parrots. They put trout in the streams so they could fish, and deer in the forests so they could hunt. They introduced weasels to keep down the rats and mice, and cats for their firesides, which left the towns and villages and took up an independent life in the countryside as hunters.

In the face of this massive invasion, the native creatures went into retreat. Those birds that had become flightless were quick to suffer. They could not escape the predatory cats and weasels, and they had lost the habit of building their nests in trees where their eggs and chicks might be safe from rats. The takahe was already close to extinction when Europeans arrived. Indeed it was first scientifically identified from semi-fossil bones. One or two living individuals were sighted in the nineteenth century, but by 1900 the species was officially declared extinct. Then, miraculously, in 1948, a small population was discovered in a remote valley in South Island. Two hundred birds are thought to survive there still, but even though they are now rigorously protected, their continued existence is still uncertain.

The flightless parrot, the kakapo, is in even greater danger. Not only were they killed by cats and weasels, but deer were eating the leaves and berries on which they depended and their numbers have fallen even lower than those of the takahé. Now a small islet, Little Barrier, has been cleared of the wild cats that once infested it and the few kakapo that survived on South Island are being collected together and re-settled in the predator-free environment that they so clearly need.

But flightless birds were not the only ones to suffer. Many others, that are able to fly perfectly well, have become greatly reduced in numbers. There were once three kinds of wattle birds on the islands. They have characteristics that link them with birds of paradise or starlings, but they are sufficiently different to be allocated a family of their own. Each has a wattle, yellow, or in one form blue, growing from the angle of the beak. One, the huia, had a sexual difference in its bill. The male's was short and used for chiselling into a tree trunk in search of grubs, while the female's was long and curved, enabling her to probe deep down their tunnels. A pair, it seems, would often collaborate, charmingly, in collecting food. The huia became extinct in the first decade of this century. Another wattle bird, the saddleback, once widespread, now survives only on offshore islands and is very rare. Only a third, the kokako, exists in any numbers on the mainland and then only on North Island. Nor is this vulnerability restricted to birds. The tuatara now is found only on offshore islands. Wetas, giant flightless grasshoppers with a savage bite and an intimidating aggressive display, are increasingly rare. The local fish, of which there were once some thirty species, have surrendered many of their streams and lakes to trout and other newcomers.

A similar fate has overtaken the inhabitants of almost every island in the world that has developed its own unique community. Exactly why this is so is still not fully understood. No doubt there are different explanations for different cases. But you might think that many island species would be so well-suited to their particular environment, and exploit it with such efficiency, that no intruder could displace them. It is not so. It is almost as though, protected by their isolation from the hurly-burly of living in a large and cosmopolitan community, islanders lose the habit of disputation and cannot maintain their position in the face of new competition. Once the barrier that protected an island is broken, it seems, many of its inhabitants are doomed.

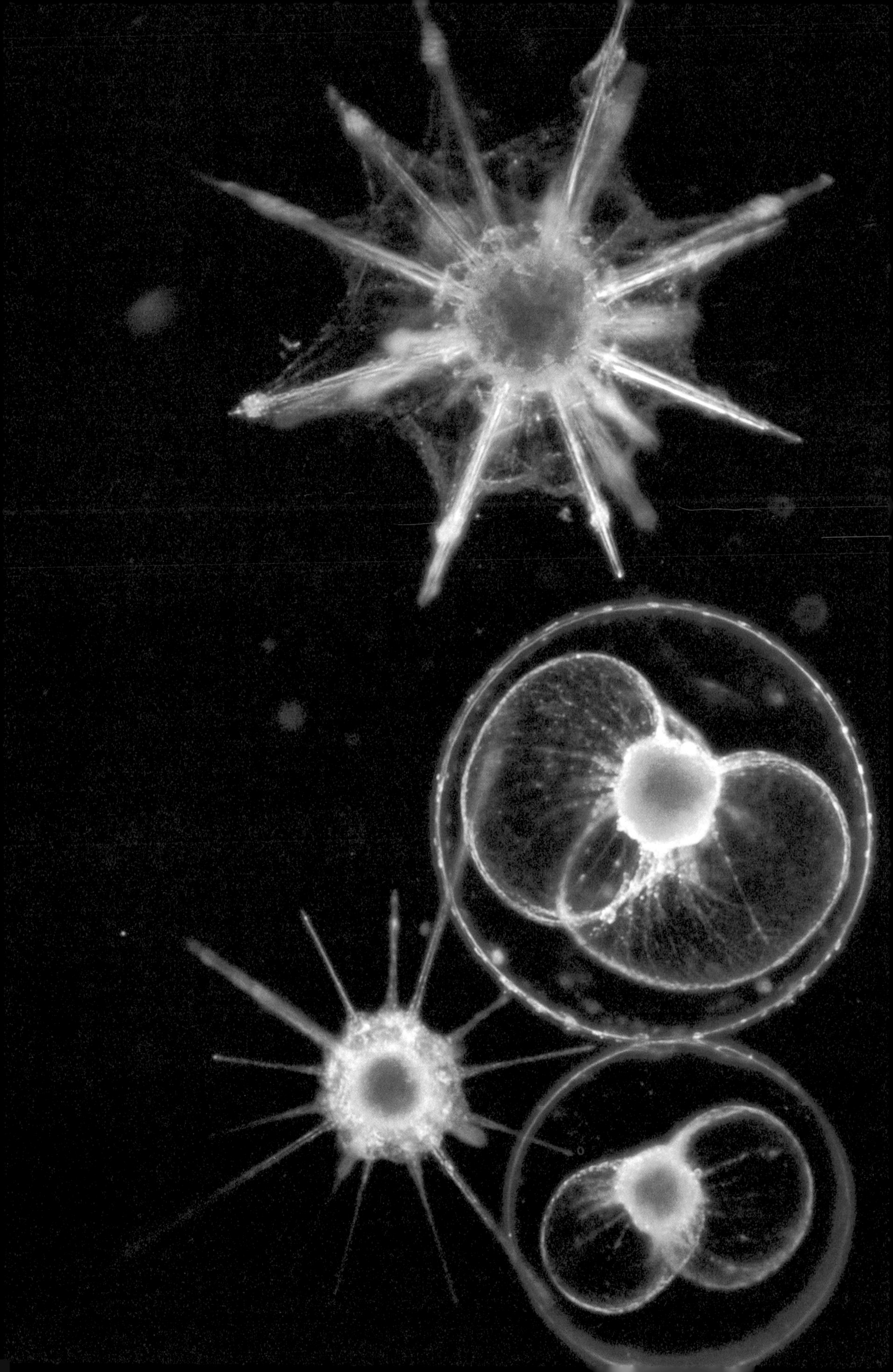

ELEVEN

THE OPEN OCEAN

Most of our planet is covered by water. There is so much of it that if all the mountains of the world were levelled and their debris dumped into the oceans, then the surface of the globe would be entirely submerged beneath water to a depth of several thousand metres. The great basins between the continents, in which all this water lies, are themselves more varied, topographically, than the surface of the land. The highest terrestrial mountain, Everest, would fit into the deepest part of the ocean, the Marianas Trench, with its peak a kilometre beneath the surface. On the other hand, the biggest mountains in the sea are so huge that they rise above the surface of the waters to form chains of islands. Mauna Kea, the highest of the Hawaiian volcanoes, measured from its base on the ocean floor, is more than 10,000 metres high and so can claim to be the highest mountain on the planet.

The seas first formed when the earth began to cool soon after its birth and hot water vapour condensed on its surface. They were further fed by water gushing through volcanic vents from the interior of the earth. The water of these young seas was not pure, like rainwater, but contained significant quantities of chlorine, bromine, iodine, boron and nitrogen as well as traces of many rarer substances. Since then, other ingredients have been added. As continental rocks weather and erode, they produce salts which are carried in solution down to the sea by the rivers. So, over millennia, the sea has been getting saltier and saltier.

Life first appeared in this chemically rich water some 3500 million years ago. We know from fossils that the first organisms were simple single-celled bacteria and algae. Organisms very like them still exist in the sea today. They are the basis of all marine life. Indeed, were it not for these algae, the seas would still be completely sterile and the land uncolonised. The biggest of them is about a millimetre across, the smallest about one-fiftieth of that. Their tiny bodies are encased in delicate shells, some of calcium carbonate, some of glassy silica. They have a multitude of exquisite shapes constructed from prongs and spears, radiating spines and delicate lattices. Some resemble minuscule sea shells, others look like flasks, pill boxes or baroque helmets. They exist in immense numbers – a cubic metre of sea water may contain 200,000 – and since they do not

Phytoplankton

propel themselves through the water but drift, they are called phytoplankton, a Greek-derived word that means, simply, wandering plants. It is these organisms that harness the energy of the sun to build, from the simple chemical substances in sea water, the complex molecules which form their tissues. So they convert mineral into vegetable.

Among them float vast numbers of small animals, the zooplankton. A large proportion of these are single-celled, like the floating algae, and differ from them primarily in the fact that they lack chlorophyll and so cannot photosynthesise for themselves. Instead, they eat those that do. There are other bigger creatures, too, of many different kinds – transparent worms studded with phosphorescent lights, small jellyfish joined together in a single rope-shaped colony a metre long, flatworms that undulate through the water, swimming crabs and vast numbers of small shrimps. All these are permanent members of this community. Others are temporary visitors – the larvae of crabs, starfish, worms and molluscs. These bear no resemblance to their adult forms but are minute transparent globes banded with lines of waving cilia. All these varied creatures feed voraciously on the floating algae or on one another, and the entire assemblage, known simply as the plankton, forms a living soup which is the staple diet of a multitude of other bigger creatures.

Plankton-feeders in shallow waters can fix themselves to the sea floor and rely on the tides and currents to bring their food to them. Sea anemones and coral polyps grope for it with their cilia-lined tentacles; barnacles snatch at it with feathery arms; and giant clams and sea-squirts filter it out by sucking the laden water through their bodies.

In mid-ocean, however, the sea floor is below the reach of sunlight and so beyond the domain of the plankton. Plankton-feeders, therefore, cannot remain attached to the bottom here, but must be active swimmers. They need not, however, swim very fast. Indeed, speed may be a waste of energy since there is a limit to the rate at which a really large collecting net can profitably pass through the water. Faster than that, a pressure ridge builds up in front of the net which deflects the oncoming water. But though plankton-feeders do not move swiftly, their diet is so nourishing that they sometimes grow to enormous size.

The manta ray, a gigantic diamond-shaped fish, grows to a width of 6 metres from fin-tip to fin-tip. It has a pair of flipper-like palps on either side of its head which channel the water into its vast rectangular mouth. The water leaves its throat through slits on either side of its head lined with combs, and these catch the plankton. A distant cousin of the ray, the basking shark, uses the same sort of apparatus to gather the same sort of food. It grows even bigger than the manta, to a length of 12 metres and a weight of 4 tonnes, and it can process 1000 tonnes of water in an hour. Its top speed is around 5 kilometres an hour, so slow that people encountering it thought it was merely lazing in the sunlit waters, 'basking', and did not realise that it was as busy as it ever gets, collecting its food.

The basking shark inhabits the colder waters of the world. Its counterpart in warmer seas is even bigger. Indeed, it is the biggest of all fish, the whale shark. This mountain of a creature is said to grow to a length of over 18 metres and a weight of at least 40 tonnes.

Zooplankton

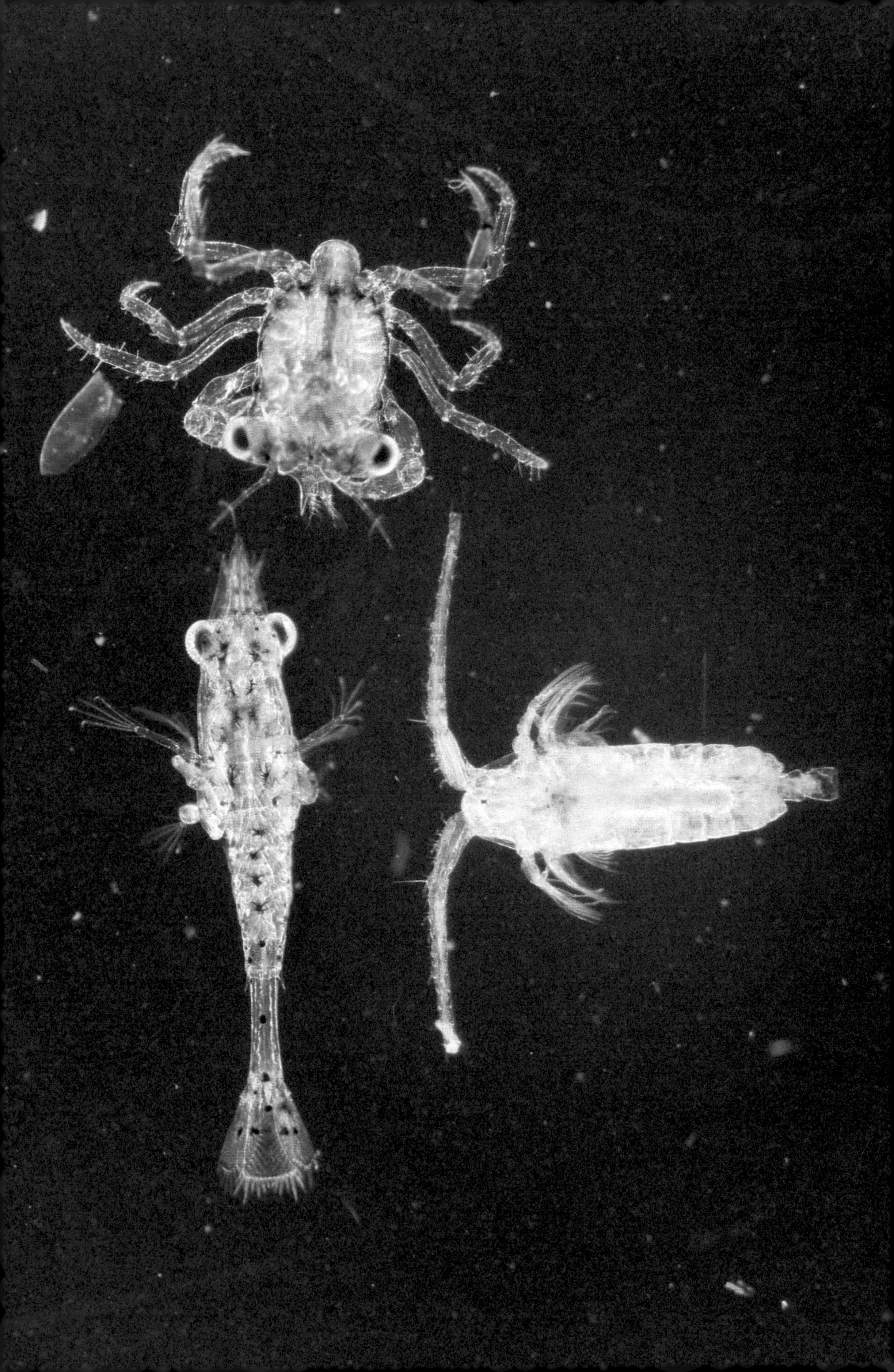

It is only rarely encountered, browsing quietly through the surface waters of the open ocean, but its immense size, its lack of speed and its harmless nature seem never to fail to make a deep impression on all lucky enough to encounter it. Occasionally one has been rammed accidentally by a ship and has hung pinned to the bows by the water until the ship has stopped, when the vast broken body has slowly detached itself and sunk to the depths. But the most marvellous encounters must be those made by underwater swimmers who have, with rare good fortune, come across one, or indeed several, for the whale shark often travels in small groups. The gigantic fish takes little or no notice of human observers as they swim around its immense body or join the attendant squadrons of fish that permanently accompany it, waiting around its mouth to pick off any morsels that stick to its tiny teeth, loitering around its tail to work through its excrement for anything edible. And then, perhaps losing patience with its new attendant, the fish inclines its great body and, with a swirl of its tail, glides down into the depths.

Manta rays, basking sharks and whale sharks belong to an ancient group of fish, the elasmobranchs, which have skeletons made of cartilage, a gristly substance, softer and more resilient than bone. By the time they arrived, all the invertebrate groups that live in the seas today had already appeared. So the early elasmobranchs had a wide variety of animal food available to them. Today, of course, the commonest members of the group, the sharks, are among the most voracious and savage of marine hunters.

Even so, we tend to exaggerate the dangers they present to man. Although some species, like the great white shark which commonly grows to 6 metres long and occasionally to almost double that, undoubtedly attack man or any other creature in the oceans, many of the smaller sharks seek much smaller prey. In the Maldive Islands, 2-metre sharks haunt the reefs and have become so accustomed to human divers that it is possible, in one place, to swim down and sit on the sea floor, some 15 metres below the surface, and watch them at close quarters. As they materialise out of the blue distance, your first reaction is not one of fear but of wonder at the sheer perfection of form of these creatures. Every contour of their bodies, every curve of their fins seem to be hydrodynamically ideal. Nothing impedes the smoothness of their passage through the water. But they do have some limitations. The pair of fins just behind the head are fixed and cannot be swivelled. So sharks have no brakes. Since they are also heavier than water they are unable, perhaps fortunately, to hover in front of the diver, nibbling him experimentally, but must either take a bite straight away or simply swoop past. And since a human swimmer is more or less their own size and much bigger than their normal meal, the Maldive sharks simply swim away with nothing more than their curiosity satisfied.

Soon after the elasmobranch lineage was founded, another kind of fish also developed from the ancient stock. Its skeleton was not of cartilage but of bone, and eventually it developed two aids to swimming that the elasmobranchs lack – an air bladder within its body that gave it buoyancy and allowed it to swim with ease at any depth it chose; and pairs of fins, fore and aft, which could be swivelled in almost any direction and so gave it great manoeuvrability in the water.

Jewel sea anemones

Whale shark

Some descendants of these early bony fish also became plankton-feeders. None of them has grown as big as their elasmobranch equivalents. Instead they have exploited the plankton riches in a different way. They form immense shoals which move and feed as single coordinated entities. Viewed in that way, the plankton-feeders among bony fish can be reckoned as having outdone even the monstrous whale shark, for such shoals are sometimes several miles across, with their individual members so tightly packed together that the centre of the shoal breaks the surface of the water like a broad wriggling hump. Anchovies operate in this way, feeding primarily on the phytoplankton. Herrings consume not only the algae but much of the zooplankton as well. Other bony fish have, like the sharks, become hunters and today there are some 20,000 members of the group which, between them, exploit virtually every environment and every food source that the oceans have to offer.

The fishes' supremacy of the sea has not, however, gone unchallenged. Some 200 million years ago, when both bony and cartilaginous fish were already well-developed and numerous, some of the cold-blooded creatures that, by then, had developed four legs and colonised the land began to return to the sea. The reptiles were the first to do so when they produced the early turtles. Later on, several groups of sea-going birds independently abandoned flying and settled down on the water. Penguins, today, are as fast and as agile in the water as many fish – as indeed they have to be, since they depend on fish for their food.

Some 150 million years ago, the mammals appeared on land, warm-blooded and covered with hair, and eventually they too produced representatives that were lured by the riches of the sea to take up residence there. The first to go, about 50 million years ago, were the ancestors of the whales. Two quite different groups of whales survive – those with teeth, like the sperm whale, dolphins and the beluga or white whale, and those with palisades of horny baleen hanging down from their upper jaw, which compete with basking sharks for larger elements of the zooplankton such as krill.

Several million years after them, another group of mammals, related perhaps to bears or otters, began to invade the sea. They gave rise eventually to today's seals, sea-lions and walruses. These creatures have not yet become as fully adapted to their sea-going life as have the whales. They still retain their hind legs, which the whales have lost, their skulls are still recognisably like those of land-living carnivores, and they have not yet acquired the knack of mating and giving birth at sea, like whales, but have to return to land every year to do so.

This procession of mammals into the sea has not yet ceased, it seems. The polar bear of the Arctic spends most of its time at sea, either on ice floes or in the water, hunting seals. It is still visibly a land animal, closely similar, except for its colour, to its near-relative the grizzly bear, but already it has developed the ability to keep its eyes open and close its nostrils under water and can stay submerged for two minutes. Maybe it, too, is starting down an evolutionary path that, if it is not interrupted, could lead its descendants in a few million years time to a fully marine existence.

So, in the 600 million years between the first appearance of multi-celled creatures

and the present day, the sea has acquired a vast and varied population of animals. All the major groups of the animal kingdom now have representatives living there. Even that most terrestrial group of creatures, the insects, has one member that lives in the sea – a water-skater that flickers across the surface of the waves. The great majority of molluscs and crustaceans and segmented worms still live in water. Many large groups – the starfish and sea urchins, jellyfish and corals, squid and octopus, as well as fish – are unable to survive for any length of time out of it. The ocean was the birthplace and the nursery of life and is still its main residence.

Just as the land contains a multitude of different environments, each with its own community of specially-adapted animals and plants, so does the sea, and there are many surprisingly close parallels that can be drawn between the two.

The tropical rain forest, the place on land where life proliferates at its most diverse and dense, has its marine equivalent in the coral reef. Superficially, the resemblance is obvious. The groves of varied corals, some with trunks and branches reaching up towards the light, others with horizontal plates catching the sunshine, look like plants. And this resemblance is closer than is often supposed.

The coral polyps which build the reef are of course, animals, not dissimilar to small sea anemones. But their bodies contain great numbers of small yellow-brown granules and these are plants, tiny algae closely related to those that swarm in the plankton. Within the polyp, these algae serve their host by absorbing its waste. They convert its phosphates and nitrates into proteins and, with the essential aid of the sun, use the carbon dioxide to produce carbohydrates. In the course of this last process they themselves release, as waste, oxygen, which is exactly what the polyp itself needs in order to breathe. So the arrangement suits both organisms admirably. In addition to these algae within the polyps, many others live independently on the dead parts of the coral colony. All in all, three-quarters of the living tissue in a clump of coral is vegetable.

Limestone, which both corals and independent algae derive so assiduously from the sea water, forms the main substance of the reef. The coral polyps, which are the major contributors, never cease secreting it. When each has built its tiny protective chamber, it sends out filaments from which sprout another polyp. This then begins construction on its own account, on top of its begetter, which, being buried, dies. So the colony as a whole consists of a thin living skin on top of layer upon layer of empty limestone chambers. This great mass of abandoned property may be dead and lifeless but it continues to serve the colony that created it by giving it massive support. To that extent, it can be compared to the inert wood in the trunk of a tree. Because the coral's algae depend on the sun, coral cannot grow much below a depth of 150 metres. And that, as it happens, is about the depth of the jungle from canopy to floor.

A great variety of creatures either feed on or make their homes in the stony thickets and branches of the reef. Parrot fish have sharp beak-like teeth at the front of their mouths with which they nip off pieces of coral, and round ones at the back to grind up the gritty mouthfuls and extract the polyps. Other fish have more delicate ways of plundering the coral. The leather-jacket, bright green with orange spots, clamps its

Beluga

mouth around the entrance to a polyp's chamber and sucks out the occupant. Starfish produce a digestive fluid which they squirt into the little compartments and so extract the polyps as a soup.

Other creatures use the reef as a place in which to hide or build their homes. Barnacles and clams bore holes into the limestone in which they can lie in safety while filtering plankton. Sea lilies and brittle stars, bristle worms and shell-less molluscs clamber continuously around the network of branches. Moray eels lurk in small caverns, ready to shoot out and claim an unsuspecting victim. Shoals of small bright-blue damsel fish, like flocks of birds, haunt the branches of antler coral, hovering just above their chosen tuft while collecting small particles of organic food from the swirling water and diving, in an instant, into the safety of the stony fronds whenever danger threatens. And around and among and between the coral colonies, packed tightly together like plants growing on the branches of a jungle tree, are sponges and sea fans, anemones and sea cucumbers, sea squirts and clams.

The diversity of rain-forest organisms, as we saw, is due partly to the excellent environmental conditions – a warm humid atmosphere and abundant sunshine – and partly to stability over a great period of time which has given evolution the opportunity to shape species to fit into a multitude of specialised niches. The superabundance of life on the coral reef is due to similar factors. The waves breaking regularly on the reef, sousing back and forth through the coral heads, saturate the water with oxygen and the tropical sun provides, throughout the year, abundant light. Furthermore, the reef is an even older environment than the rain forest. Reefs containing species of coral, sea urchins, brittle stars, molluscs and sponges, all closely related to species found on the reefs today, were well established some 200 million years ago, as their numerous fossils make clear. Between then and now, there have always been reefs in some part of the tropical seas, always places for the ubiquitous larvae in the plankton to find sites to settle. Today, the Great Barrier Reef off eastern Australia, to take one example, contains over 3000 different species of animal, and most of them in huge numbers.

This dense crowding brings its own problems. Any hole or cranny that can offer reasonable shelter is vigorously contested. One kind of shrimp habitually and laboriously excavates a hole for itself in the sand between the coral heads; and, equally habitually, one kind of blenny moves in alongside the shrimp and uses the hole as its own refuge. Empty mollusc shells are occupied inside by hermit crabs and outside by sponges. These flourish on the crumbs of the crab's meals and envelop the shell so completely that the crab's refuge is invisible to predators. The pearlfish, as long and as slim as a pencil, finds its shelter inside another animal's body. It gains entrance to the interior of a sea cucumber by nudging the cucumber's anus with its nose. Once inside, it is not only protected from its enemies, but it has a readily-available food supply. It nibbles the cucumber's internal organs and the cucumber obligingly regrows them as fast as it loses them.

Crowding may also explain the splendour of many reef creatures. An individual fish, here as elsewhere, must be able to recognise which among the swarms of fish around it

Butterfly fish feeding among coral

Overleaf: *Jack feeding on fry*

are its own species and therefore potential mates or rivals. In the visual clamour of the reef, identification signals have to be particularly vivid if they are to register. The problem is acute when several related species, similar in shape and size but each cropping its own particular food source, swim in the same waters. The butterfly fish are just such a family, and each species has its characteristic and often extraordinarily beautiful combination of eye-spots, bars, patches and dots, so that, like gorgeous butterflies in the jungle, each can be recognised at a distance.

If coral reefs are the jungles of the sea, then the surface waters of the open ocean must be its savannahs and plains. There, year after year, over vast areas, the phytoplankton blooms. Like grass, its abundance varies with the season, for, like all plants, it requires not only light but phosphates, nitrates and other nutrients. These come from the droppings and dead bodies of the multitude of creatures that live on the surface. But unlike the droppings of cows in a field, they do not remain on the pasture. Instead they fall steadily and gently down through the water to accumulate as an ooze on the sea floor, far beyond the reach of the floating algae. But when seasonal storms stir the seas, the fertilising ooze swirls upwards. Suddenly the phytoplankton can grow again and it does so with great vigour. By the time the calmer months have passed, the algae have flourished and increased so greatly that they have exhausted most of their chemical food and the waters are once again impoverished. So the plankton dies back and will remain at a low level until the annual storms once more refertilise the water.

The shoals of anchovy and herring, sardine and flying fish that graze these vast meadows are hunted by packs of voracious, carnivorous fish, just as herds of antelope on the plains of Africa are preyed upon by cheetah and lion. Some of these marine hunters, like mackerel, are not greatly larger than their prey. Others, such as the 2-metre-long barracuda, take not only the plankton-feeders, but also the smaller hunters. Biggest of all are the great sharks and that group of magnificent ocean-going fish, the tuna. Both shark and tuna are high-speed swimmers, as they have to be if they are to catch their prey. They reach a comparable size and have very similar body shapes, but it is the tuna and their near relatives, the billfish, that most closely approach swimming perfection.

These superb creatures roam the seas worldwide. There are some thirty different species of them and they have acquired a variety of names from fishermen of different nationalities who search eagerly for them, not only because of their rich flesh but because they are such valiant and powerful fighters. Among them are tunny and albacore, king mackerel and skipjack, marlin and sailfish, bonito and wahoo. Some of them grow as big as 4 metres and as heavy as 650 kilos. A gigantic swordfish has been caught which measured 6 metres long, making its species the largest of all the bony fish. The shape of them all is hydrodynamically even closer to the ideal than that of the sharks. The snout is pointed and sometimes extended into a long sharp spike, like the nose of a supersonic plane. The rear part of the body tapers gently and ends in a crescent-shaped tail. The surface of the eyes is contoured so that it does not bulge or interrupt the smooth streamlining of the head. Tuna, and some others, have a corselet of specially modified

Garden eel

scales just behind the head that serves as a spoiler, producing a slight turbulence around the widest part of the body which reduces the drag on the hinder end. When swimming at speed, the fins slot into special grooves so that they do not impede the flow of water. One of these superb fish, the sailfish, holds the speed record for any marine creature and has been timed, swimming over a short distance, at 110 kilometres an hour. That is even faster than the cheetah, the fastest of all land animals.

Swimming at such speeds consumes a great deal of energy and therefore demands an abundant supply of oxygen. These fish obtain it, not by lowering the floor of the throat and pumping water gently into their mouths and out through the gill slits by moving their gill covers, but by swimming with their mouths permanently open, forcing a high-speed jet of water over their large gills. They have, as a consequence, to swim continuously at a considerable speed simply in order to breathe. The energy output of their muscles and their lightning reactions are also enhanced by maintaining the chemistry of their bodies at a high temperature. These fish, in fact, unlike any others, are warm-blooded, with body temperatures that may be as much as 12°C above the waters through which they swim.

Swordfish are usually solitary hunters. They dive into the shoals, slashing with their long rapiers, sometimes, it is said, stabbing some of their prey and stunning others. Tuna usually operate in squadrons. A group has been observed carefully herding a shoal, driving it from behind and patrolling its flanks to keep it well-packed. When they attack, there is wholesale carnage. As they rip through the shoal, snatching the small fish with devastating precision and speed, the shoal itself panics. Fish shoot from the surface of the sea in hundreds in an attempt to escape the snapping jaws below, like terrified impala leaping away from a team of rampaging lions.

As well as its savannahs, the sea has its deserts. Near the edges of the continents, huge expanses of sand cover the sea floor. Compared with the surface waters, this part of the sea seems virtually lifeless. Currents, sweeping across the sands, blow them up into long ripples and dunes, just as winds do in a land desert. Sand in itself contains no nutrients, and such organic particles as are deposited among the grains are winnowed away again by the currents that continually sift and shift them. Some creatures manage to live here as they do on the sandy beaches closer to the shore. The garden eel buries its tail in the sand, secretes a mucus which holds the sand grains together, and then rears upright with the top part of its body in the open water, filter-feeding. One species of anemone, like the sand-mason worm, constructs a free-standing tube of sand for itself. Such creatures as these seem, at first sight, to be almost the only inhabitants of these otherwise deserted parts of the sea floor. But that is an illusion. A great number of animals live within the sand itself. Lying close to the surface, covered by a light dusting of sand grains to perfect their camouflage, are flat-fish – plaice, sole, skate and halibut. Buried deeper in it are hosts of invertebrate animals of many kinds – molluscs, worms and sea urchins.

One part of the ocean, however, has no parallel whatever on land. Beyond the sandy deserts around its margins, beneath the planktonic meadows near its surface, lie the

Deep-sea angler fish with two parasitic males

black depths. Until recently, our knowledge of what lives there was based almost entirely on the largely haphazard selection of mangled corpses brought up by deep-sea dredges. Now, however, there are a number of craft that can descend several kilometres below the surface of the sea and the beams of their searchlights have given us glimpses of a world more remote from us in space and physical conditions than any other part of the planet, above or below water.

As you descend, the water becomes colder and colder. Soon it is close to freezing. Beyond 600 metres, the light of the sun has been totally obliterated by the mass of water above. Every 10 metres of descent increases the pressure by one atmosphere so that at 3000 metres, it is about 300 times the pressure of air at the surface. Food is very scarce indeed. The dead bodies that drift down from above fall very slowly. A small shrimp may take a week to reach 3000 metres. Consequently, most have been eaten before they reach such a depth or have decomposed so much that they are beyond the absorptive powers of any animal gut and are fit only for fertilising the plankton. And yet, even our limited exploration of this remote world has already revealed over 2000 species of fish and a similar number of invertebrates.

Over half of them provide their own light. The batteries they use are, in nearly all cases, colonies of bacteria which glow as a by-product of their own chemistry. Fish maintain colonies of them in special pouches, on the side of the head, the flanks, or at the end of a fin ray. The bacteria themselves shine continuously but this may not suit their owners who may well, on occasion, greatly prefer invisibility. So the fish turn off their bacterial torches by raising opaque shutters of tissue in front of them, or restricting the supply of blood to them.

So many fish in the middle and deep waters of the ocean are luminous that it is clear that the possession of light is very important. But there is still a great deal to be learned about the exact purpose it serves. The little flashlight fish keeps its bacteria in a small chamber below each eye. They swim in shoals switching their lights on and off by raising a small screen of skin in front of them. Presumably the particular semaphore of the flashing enables the shoal to keep together and male to find female. If a predator approaches and the shoal is alarmed, they all switch off their lights and swim rapidly away – and then begin flashing at one another somewhere else. A great number of fish carry their lights on the underside of their bodies, suggesting that the creatures for which they are intended must be somewhere in the water below them. It may be that these lights have some value, paradoxically, as camouflage. In those upper parts of the deep sea where light from the distant surface is just bright enough for a fish to form a silhouette against it, an illuminated underside may have the effect of making it less visible.

Such functions may sound unlikely, and certainly there is a great deal we do not yet understand. But it is beyond question that light does serve as an attraction in the blackness and some fish use it as a means of luring their prey within range. Angler fish in shallower seas attract their prey with a long, specially modified dorsal spine dangling in front of their mouth. This has a tiny membrane at the end of it which flutters as the

Capelin and their spawn, Newfoundland

fish shakes it like a fisherman's lure – which is exactly what it is. In deep-sea anglers, this lure is a bulb of bacterial light. Small fish seem irresistibly attracted to it. They swim closer and closer until eventually the angler sucks them into its jaws.

The need to attract prey is especially important, for although there are a great number of species in the deep sea, the density of individuals is very low. Encounters, therefore, are few and when they do take place full advantage must be taken of them. This may explain why so many of the deep-sea fish have gigantic distendable bellies capable of accommodating prey substantially bigger than the hunter was itself before it tackled its meal. It also accounts for the strange sexual relationships of many deep-sea anglers. When young, a male is somewhat smaller than a female, but otherwise not very different from her. If he succeeds in locating a female, he attaches himself by his jaws to a place on the female's body close to her genital opening. Then he slowly degenerates. His blood system unites with hers and his heart withers. Eventually, he becomes little more than a bag producing sperm, but he will continue fertilising her eggs for the rest of her life. He has taken advantage of his one sexual encounter to the fullest.

The very deepest parts of the ocean lie below the path of the currents so the water is not only black and cold, but very still. This also has its effect on the shape of the fish. With no current to swim against, they need few muscles with which to swim and maintain their position. This gives them a characteristically fragile appearance and makes them, since many are almost transparent, reminiscent of the fantasies of Venetian glass-blowers. It also makes it possible for those that live right on the bottom to move across it on the slenderest of stilt-like fins.

The floor of the deeps in the centre of the ocean basins lie, for the most part, far beyond the reach of land-derived sediments of any kind. The only mineral particles to drift down here are grains of volcanic dust, falling from the atmosphere. The pressure is so great that bones and limestone shells disintegrate. The skeletons of those phytoplankton that build with silica are more resistant and, oddly, so are the ear bones of whales, the jaws of squid and the teeth of sharks. The great pressure, however, does cause the water to precipitate some of its dissolved minerals and in some parts the deepest ocean floor is carpeted with nodules of manganese, iron and nickel, some as small as grapes, some as big as cannon balls. Even here, the searching lights of deep-sea craft have found signs of life – wandering trails in the meagre ooze made by worms laboriously munching their way through the sediment to extract its last edible particle.

But much of the ooze, even that part of it which derives from the dead bodies or the droppings of animals living above, is not edible. It has decomposed into its chemical constituents, such as phosphates and nitrates and these can only be reassembled into organic tissues by bacteria and plants. No alga, of course, can live in these lightless depths, so the fertilising ooze remains beyond the reach of phytoplankton until storms stir it. One other force may also produce the same result. In some places, a powerful current flows across the deep ocean floor, sweeping up the ooze and bringing it back into circulation.

One such current starts in the Caribbean Sea. This small arm of the tropical Atlantic lies warming in the sun in a relatively shallow basin, hemmed in against the eastern coast of Central America by the islands of the West Indies. The forces produced by the spinning of the earth, reinforced by unremitting trade winds, push the Caribbean waters north and west between Cuba and the peninsula of Yucatan up into the Gulf of Mexico. From there they go on, running like a vast warm river, 80 kilometres wide and 500 metres deep, carrying with them a rich burden of tropical plankton, through the colder waters of the western Atlantic up the eastern coast of America. This is the Gulf Stream. After it has travelled some 5000 kilometres it meets head-on another great river flowing southwards through the ocean from the Arctic, the Labrador Current. The warm air and the cold air travelling above the two currents mingle and produce fogs that linger throughout the year. Beneath, the waters churn and boil.

It so happens that at this particular meeting point, a large submarine plateau, 300 kilometres wide and 500 long, rises from the depths of the Atlantic. It approaches so close to the surface that all the water above it is within reach of the sun's rays, so the phytoplankton flourishes. But it never exhausts its supply of nutrients, as happens elsewhere, for the currents sweeping up the sides of the plateau from all directions scoop up the fertilising ooze from the deep sea. The result is a never-ending supply of planktonic soup of an unparalleled richness, and shoals of fish flourish here as nowhere else. These are the Grand Banks of Newfoundland.

Feeding directly on the phytoplankton are capelin, a small fish distantly related to the sardine. During the summer, they shoal in incalculable numbers off Newfoundland's sandy beaches, darkening the waters. On spring tides, they come inshore and as high water approaches, they swim up on to the beach itself. Each wave brings in thousand upon thousand. As it dumps them on the sand, the females, with a swift and urgent wriggle, dig a shallow groove and expel their eggs. The males, alongside them, deposit sperm and the next wave carries them back to the sea. But not, for most, to a new life. Nearly all of them, having spawned, die, and their pallid corpses accumulate in the shallows offshore in vast drifts.

The capelin shoals lure many other animals here. Tens of millions of cod feast upon them. Sea birds descend upon them from the skies above. Gannets plunge-dive on them, bombarding the water; kittiwakes and razorbills paddle among them. Seals cut through the turbulent waters, stuffing themselves with the tiny fish. And most impressive of all, humpbacked whales come to gulp up tens of thousands in every mouthful.

And men, too, come to harvest the wealth. Ever since industrial fishing developed, the Grand Banks have been scoured with increasing intensity. Year after year fishermen have brought new ways of locating the shoals with radar and sonar, new designs for nets, new techniques for taking away greater and greater catches. But even the Grand Banks are not inexhaustible. Today, modern fish-processing factories, built only a few years ago on the coast nearby in the belief that the vast harvest could be sustained year after year, stand idle and empty. The catches are failing. Man's greed has put at risk the survival of even the richest and most productive part of his planet.

TWELVE

NEW WORLDS

Living organisms are extraordinarily adaptable. Species, far from being fixed and immutable, evolve with a speed that is well able to match most geological and climatic change. Owls, colonising the far north, developed the thicker, whiter plumage that now keeps them warm and inconspicuous on the snow-covered tundra. Wolves, finding their habitat changing to desert or extending their territory into it, lost their thick fur, and so their bodies do not overheat. Antelope, moving out from forests and grazing on open savannahs, grew longer legs and became swifter runners, and so the hazards of living in such exposed circumstances were reduced.

Man, for the first few millennia after his appearance as a new species, showed signs of the same adaptability. Eskimos, living in the Arctic, developed short, stocky bodies, the shape that tends to retain heat; Indians in the Amazonian rain forest have hairless bodies and long thin limbs, the shape that tends to lose heat. Those people who live where the sunshine is so fierce that it can damage their bodies have dark pigmentation in their skins; those in cloudier, cooler regions where sunshine is so feeble and infrequent that it is scarcely sufficient to promote the production of vitamins in the body have less pigment and pale skins.

Then, some 12,000 years ago, mankind began to show a new talent. When faced with harsh surroundings, he no longer waited many generations for his anatomy to change. Instead, he changed his surroundings. He began to modify the land in which he lived and the animals and plants on which he depended.

The people who made one of the first moves in this direction lived in the Middle East. At this time, they were still wanderers hunting wild creatures, gathering roots and leaves, fruits and seeds. They competed for their prey with packs of wolves. Doubtless, the wolf packs followed the human hunters, picking up unwanted offal, just as jackals in Africa associate with prides of lions and collect a share of the kill when the lions have taken all they need. Maybe, too, things sometimes worked in the other direction. A pack of wolves might make a kill and the human hunters claim some of it.

The two species shared not only the same territory and the same prey, but similar social organisation. Both hunted in groups and both had complex hierarchies in which a

Mouflon

chain of authority and command was established by regular displays of dominance and submission. Eventually, the two species came to form an alliance.

It is not difficult to imagine how it happened. Tribal people everywhere take pleasure in keeping pets, so it is a reasonable guess that some of these early hunters took young wolf pups and kept them with their own children around the camp fires. Maybe nursing human mothers even took orphaned pups and gave them a share of their breast milk, as some tribal people will suckle piglets today. So young wolves, raised in the human pack, may have come to accept the dominance of a human leader. When they became full-grown that dominance remained and they joined their human master in the chase, accepting his instructions and being rewarded with a share of the kill.

Among the animals pursued by man and dog at this time were wild sheep. The mouflon, which still lives in remote parts of Europe, is probably very similar to the wild sheep of that period. It is small and long-legged. Both ram and ewe carry heavy ringed horns. In winter, it develops a woolly undercoat and in summer it sheds it. Around 8000 years ago, man developed a special relationship with this shy, nervous creature. The process must have been very different from that by which he recruited the dog and may well have been very similar to that in progress today between man and the reindeer that graze on the tundra in Northern Europe.

These animals are wanderers, for their pasturage, especially in winter, is so poor that they must move continually from one patch of country to another to find new areas of uncropped moss and dwarf juniper. They are followed by the Lapps, nomads who may have originated somewhere in central Europe and who migrated to the Arctic about 1000 years ago. These people are totally dependent upon the reindeer, which provide them with all the essentials of life – meat and milk for food, thickly furred pelts for clothes, skin stripped of fur for tents, sinews for sewing, rawhide for ropes, antlers and bone for tools. But the Lapps cannot be properly described as hunters in the normal sense, for the reindeer today are no longer truly wild.

Even though the Lapps cannot control the wanderings of the animals, individual families regard particular herds as belonging to them. The calves, which appear each spring, they also regard as their own property. There is, however, one problem. Young males tend to be driven out of the herd by the dominant stags and to wander away to establish their own groups. They could, therefore, be lost to their owners. But if they are castrated they do not challenge the dominant stags and remain with the herd. So each year the Lapps round up their animals, mark them and castrate them.

Some young males, of course, must be left unmutilated to sire future generations, and it is only sensible to select for this purpose those animals that are the most docile, and the most likely to remain with the herd, even when they are sexually active. This selection has been going on for many centuries. So, without necessarily having any particular intention to impose selective breeding, the Lapps have been practising it. Today, their reindeer are very docile and remain together throughout the year in great herds that may number more than a thousand, something that the reindeer's totally wild North American relation, the caribou, will not do.

Lapps rounding up reindeer

With such unwitting management as this, man may have eventually created his herds of obedient sheep and goats. For a thousand years or so these remained his only domesticated food-animals. Then, at last, he managed to tame cattle. This must have been a much more difficult and, indeed, dangerous process. The wild ox that wandered over Europe and the Middle East 8000 years ago was a huge animal, the aurochs. The last of the species died in the forests of Poland 300 years ago, but we know how big they were from their bones and how impressive they looked from the vivid portraits on the walls of caves in France and Spain that were drawn by hunters of a much earlier period. They stood up to 2 metres high at the shoulder. The bulls were black with a white line running down their spine, the cows and calves a little smaller and reddish-brown. They must have been very formidable animals, but men, aided by their packs of dogs, certainly hunted them and did so very effectively, for the remains of their kills and feasts have been discovered. And men did more than hunt the aurochs. They worshipped it. In the settlement of Çatal Huyuk in Turkey, built some 8,000 years ago, a room was found in which bony horn-cores of aurochs stand in lines mounted on a clay bench. It seems that it can only have been a shrine.

The veneration of wild bulls continued for a long time. Hinduism, the most ancient of all the world's great religions still reveres them. The Roman god, Mithras, was associated with them and those practising his cult had to sacrifice them. The ceremonial slaughter of bulls in an arena, still practised in Spain, may derive from the same source. As centuries passed, these sacred wild animals were also tamed and man began to breed them selectively to create a kind of cattle that suited him better. Not surprisingly, one of the first changes he brought about was a reduction in size, for beasts as big as the wild aurochs must have been very difficult to control.

Some of these early domesticated breeds still survive. In Britain, a herd was fenced in during the thirteenth century and it still lives in a large walled park at Chillingham in the Cheviot Hills. Though these cattle are small in comparison with the aurochs, the bulls are still extremely aggressive. If human beings approach them, they form in a ring, horns outwards, ready to charge any attacker from any quarter. One great bull rules the whole herd. He mates with all the cows and fights every young male who challenges him until eventually, after two or three years, he loses and surrenders his place. Today, no one tries to coerce them in any way and it is said that if a calf is even touched by human hands, the herd will kill it.

The Chillingham cattle are pure white, unlike the wild aurochs. This change may be significant, for many domesticated animals are white or piebald. As well as sheep and goats, later additions to man's society such as pigs and horses and, in the New World, llamas and guinea pigs, all have races with this conspicuous colouring. When, as a result of some genetical quirk, such individuals appear in a wild population, they are at a considerable disadvantage for they are quickly picked out by predators. Under man's protection, however, this does not happen, so the genetic tendency flourishes unrestrained and spreads among the group. Indeed, it could be that herdsmen actually preferred such vivid colouring since it enabled them to keep track of their animals browsing in the woodlands, and that they therefore, at a very early date, deliberately selected individuals that were so coloured and bred from them.

At about the same time as man was bringing animals under his control and changing their shape, he was doing the same thing with plants. Grass seeds had long been gathered for food, as they are still today by Bushmen in the Kalahari and by Australian Aborigines. Ripe seeds, however, are easier to collect when they are still attached to the seed head, rather than after they have fallen. So women, who may well have undertaken the job of gathering as they do in most foraging societies today, would have tended to select these. When people began to take to a more settled life and build permanent dwellings, the grain they saved for planting nearby would, therefore, have this characteristic. So mankind, ignorant though he may have been of any of the principles of plant breeding, began to bring into existence a new kind of grass that was easier to harvest. In order to plant it he began to clear the land around his settlements, cutting the trees and uprooting the bushes to provide space and light for his crop. Mankind had become a farmer.

These new versions of plants and animals slowly spread from settlement to settlement across the Middle East and into Europe. As they were adopted, so people changed the face of the land to accommodate them. Just how drastic and wholesale these changes eventually became can be seen vividly in Britain. Ten thousand years ago, the British Isles were covered almost entirely by woodland. In the north of England and in Scotland, there were forests of evergreen pine; in the south, mixed deciduous woodlands, dominated by oak, lime and elm, with smaller numbers of hazel, birch, alder and ash. Only swamplands and the slopes of hills above 700 metres were bare. Human beings were living in the woods and had been for many thousands of years. But so far they had altered them hardly at all. They collected hazel nuts and wild fruits and hunted, with the aid of their dogs, not only aurochs but red deer and elk, beaver, reindeer and wild pig. Then, around 5500 years ago, the farming people from Europe began to arrive in southern England. They brought with them seeds of cultivated wheat and herds of domesticated sheep and cattle. They began to cut down the forests, using stone axes, to make room for their settlements, to provide grazing for their livestock and fields for their crops of grain.

Today we are inclined to regard the landscape these people created as the very epitome of the natural English countryside – the swelling chalk downs carpeted by a close turf, golden with a haze of cowslips in spring and in summer spangled with tiny brilliantly coloured flowers, and, in the clear blue sky above, larks spilling silvery cascades of song. In fact, everything in that landscape except the underlying shape of the chalk hills themselves is there as a result of the actions of man and his animals. He cut down the trees and his animals have prevented them ever since from regenerating by nibbling every young tree seedling that germinates.

Such transformations have now affected almost every part of Britain. Man's responsibility for them, however, is often forgotten. The Norfolk Broads, that wilderness of reed-beds and waterways, are not natural lakes but vast pits left by peat-diggers in

medieval times which have subsequently flooded. The heather-covered grouse moors of the Scottish Highlands were once pine forests, and were cleared in some instances as recently as 200 years ago. Man promoted the growth of heather in their stead in order to increase the number of grouse, which feed on heather leaves, and he maintains them in this condition by systematically burning every part of the moor once every ten to fifteen years. The uniform rectangular plantations of conifers that cover the flanks of many British hills are only too obviously man-made, but even the woods and spinneys of mixed trees that add so much interest to lowland country and support such a variety of wild life were mostly planted by man in order to provide cover for game and to be cropped for their timber.

As man changed the British landscape, so he also changed the animals that lived in it. Those that did not suit him or that he considered dangerous, such as wolves and bears, he exterminated. Others, beaver, reindeer, elk, he allowed to disappear largely by accident; he overhunted them or destroyed the sort of country they needed in which to live. At the same time, he introduced others from elsewhere. In the twelfth century he brought in the rabbit, originally a native of western Mediterranean countries, for the sake of its flesh and its fur. Within a couple of centuries it had become the most numerous of all the larger quadrupeds in the country. Around the same period he also introduced pheasants, which came originally from the Caucasus. New strains of the bird have been added on several occasions since, including, notably, ring-necked pheasants from China. They too are now well-established, free-living inhabitants of the countryside. So, over the centuries, more and more creatures were added to the British community, to provide either food, sport or decoration, or all three, so that today there are at least thirteen mammals, ten birds, three amphibians and ten fish, all from foreign parts, that are naturalised inhabitants of Britain.

Man also continued to mould his domesticated creatures still further to suit his requirements. He produced sheep that bore thicker fleeces and carried them all the year round, so that the wool was not shed piecemeal but could be shorn and collected when it suited the shepherds. He bred cows that had lost nearly all of their aggressiveness, that produced unnaturally abundant supplies of milk and put on unnecessary muscle in those parts of its body that best suited the human cooks of the time. Dogs he diversified to an extraordinary degree. Mastiffs were bred to serve as baleful guardians that could bring down a man; spaniels to have a highly developed sense of smell that enabled them to retrieve birds shot from the sky; terriers, short-legged and pugnacious, to go down holes and fight foxes; long low dachshunds to hunt badgers; bulldogs with jutting lower jaws and over-developed fangs that would grip a baited bull and hang on, no matter what blows they were given; and, surprisingly early, soft-haired, large-eyed dogs that remained puppyish throughout their lives and would sit on a lady's lap to be petted. Though all these breeds were derived from the same wolf ancestors, some became effectively new species, the most extreme forms being unable to interbreed with others for simple reasons of proportion and stature.

He treated plants in the same way. Our kitchen gardens today contain vegetables

Sheep-grazed turf, South Downs of England

from all over the world. Potatoes were first cultivated by the Incas in the Andes; runner beans, sweet corn and tomatoes by the Aztecs in Mexico. Rhubarb came from China, carrots from Afghanistan, cauliflowers from the Middle East and spinach from Persia. And all, over the past 500 years, have been bred into varieties that greatly exaggerate the parts of them that we most value for food, and in some cases been transformed almost beyond recognition.

And man also created a completely new environment. He built towns. The first sprang up in the Middle East some 10,000 years ago and seems to have been directly connected with the first domestication of plants and animals that freed man from the necessity to travel in search of his foods. These dense settlements, accommodating several thousand people, were made of sun-dried mud bricks and, doubtless, initially were not such alien places. Plants must have been able to get a roothold without much difficulty in the crumbling brickwork. There were plenty of dusty corners where spiders could spin their webs, and piles of rubbish where field mice could hide and nest. But as man's technical skills developed, as he learned to build with more durable materials such as stone and kiln-fired bricks, as he paved and cobbled his roads, so his towns became less and less hospitable to the creatures from the wild. Today, man has become so ingenious as an engineer, so inventive as a producer of new materials, that there is often little in his towns that is not man-made. It is hardly possible to imagine any environment more divorced from the natural world than that represented by the Sears Building in Chicago. It is currently the tallest building ever constructed, standing 450 metres high. Its skeleton is of steel girders, its outer surface a polished vertical precipice of bronze-faced glass, black-skinned aluminium and stainless steel. Twelve thousand people migrate to it every morning and spend their days within it, most far beyond the reach of the sun and breathing air, purified, humidified and at the right temperature for their comfort, that is delivered to them by computer-controlled pumps. For many miles around, the soil of the land has been sealed beneath asphalt and concrete, the air filled with gases from the exhaust of cars and the breath of a million air-conditioners. You might suppose that such cities would have no place in them for any forms of life other than human beings. Yet plants and animals have responded to this new environment as they have to all others on the surface of the earth; they have discovered not only how to tolerate these new circumstances but, in some instances, have come to prefer them to all others.

The bleak sterility of masonry and concrete does, in fact, have one natural counterpart – the ash fields and lava flows of volcanoes. Plants that evolved to colonise one are sometimes able to colonise the other. Back in the eighteenth century, an Oxford botanist collected from the slopes of Mount Etna in Sicily a tall daisy-like plant with bright yellow flowers and brought it back to the University's botanic gardens. There it flourished so well that by the end of the century it had escaped and was growing on the limestone walls of the colleges. For several decades, it made no further move. But in the middle of the nineteenth century, railways were being constructed across the country and the embankments and cuttings, liberally sprayed with cinders and ash from the locomotives

Sears building, Chicago

that ran through them, proved much to its liking. Soon the Oxford ragwort, as it was now called, was travelling along the lines and reaching new territory. Today, there is scarcely a city in Britain where you cannot find it growing on a vacant building site among the tumbled masonry.

The fire-weed that colonises the flanks of North American volcanoes and is currently reclaiming the ash-covered slopes of Mount St Helen's also has a similar chapter in its history. It was regarded as an uncommon species in Britain during the last century. But when great areas of British cities were laid waste by bombing during the Second World War, the fire-weed suddenly proliferated, cloaking the ruins with dense sheets of purple. Now, it has become one of the commonest of wild urban plants in Britain where it is known as rosebay willow-herb.

Animals, too, have been able to find man-made equivalents to their natural homes. The vertical faces of buildings can provide, in the hands of some architects, very much the same nesting opportunities as cliff faces, so birds that frequent such habitats have no difficulty in taking up an urban existence. One of the commonest and most typical of city birds, the pigeon, is a descendant of the rock dove that originally lived on sea cliffs and today survives in Britain in its unchanged form only in Ireland and parts of Scotland. This dove was domesticated some 5000 years ago for the sake of its flesh and provided with special dovecotes in which to live and nest. But it has since returned to a free existence in cities, there to be joined by genuine wild birds. The two have now interbred to produce the variegated flocks that fill the skies above almost every public place in the cities of western Europe. Among them are some individuals that resemble the original wild rock dove very closely indeed, with blue-grey plumage, white rump, and glossy purplish-green on the head and neck. They differ only in that the band of bare skin at the base of the bill is a little more prominent. Others carry the characteristics isolated and emphasised by many centuries of domestic breeding, being white, black, piebald or terracotta. Urban pigeons build their nests among classical capitals and neo-Gothic niches just as they would on the ledges and crevices of a sea cliff. Starlings assemble in towns in autumn in flocks of tens of thousands to roost in the buildings and benefit from the warmth which, on cold days, may be several degrees above the surrounding countryside. Kestrels live in spires and turrets, surveying the ground below for prey as their country cousins do from rocky pinnacles. Many houses have dark lofts and attics immediately beneath their roofs which can be reached through a gap left by a dislodged brick or slate. Bats find such places just as convenient for their roosts as a cave. In North America, a swift that originally built its ledge-like nest on the inside of hollow trees, found that in many parts of its territory there were more ventilator shafts and chimneys than there were hollow trees. Today, the chimney swift nests hardly anywhere else but in towns. In tropical cities, vertical concrete walls and window panes are ideal territory for lizards whose talent for adhesion allows them to run about on smooth leaves and vertical tree trunks. So now few houses in the tropical Far East are without their populations of geckos, attentively snapping up insects that are attracted indoors by the artificial lights.

Oxford ragwort

Some of these urban immigrants have found concentrations of exactly the food they most favour. The grubs of some moths get fat by munching their way through piles of woollen clothes. Weevils devastate grain stores if once they gain entry, feeding and breeding continuously until they consume and contaminate all the grain within reach. Termites and beetle grubs burrow in the timber of joists and furniture. Some termites have even developed a taste for plastic, stripping cables and causing major electrical faults, though it is difficult to understand what attracts them since the plastic they chew so assiduously has no discoverable nutritional value whatsoever. Perhaps they, like chewers of gum, find the activity rewarding simply in itself.

The great majority of urban animals, however, have been attracted to the city by one great food resource – man's waste. The discarded take-away meal, the carelessly spilled crumbs, the garbage can and the refuse tip, these are the urban equivalent of plankton in the ocean, or grass on the savannah. They provide the nutritional base on which whole chains of animals, one feeding on the other, are founded. Predominant among the consumers of this waste are rodents.

The house mouse is not the same species as the field mouse, which hardly ever ventures far into towns. Where it came from is difficult to determine, but it may have lived somewhere in the semi-deserts of the Middle East or perhaps the steppes of central Asia. It joined man soon after he built his first settlements and it has been with him ever since, following him all over the world. Basically, all house mice everywhere belong to the same species, though it is possible to identify different races among them. Populations within cities form isolated communities, cut off as they are from other cities by barriers of countryside. Evolution in these urban islands proceeds at a particularly swift rate, as it does on islands and in lakes, perpetuating tiny inconsequential differences in anatomy and even, on occasion, producing special adaptations. So several big cities in South America have their own identifiable races of house mice and some long-established refrigerated cold stores have now developed their own dynasties of resident mice which have specially thick fur that keeps them warm in the Arctic conditions.

The black rat also joined man at a very early stage. It lived, somewhere in southeast Asia, in trees, and has never lost its predilection for climbing, finding itself very much at home on ships, particularly wooden sailing ships, where it scampers up and down the rigging with agility. This liking for ships led to its rapid spread around the world. Black rats were abundant in the cities of continental Europe by the twelfth century and they reached Britain soon after, carried, it is said, on the ships of returning Crusaders. By the middle of the sixteenth century, they had managed to get passage across the Atlantic and had appeared in the cities of South America.

The brown rat joined man rather later. It also originated in Asia, but it was a burrower rather than a climber. It too has retained its ancestral preference, so that where brown and black rats infest the same building, the black lives in the upper storeys, running along pipes and rafters, while the brown gnaws holes in the wainscot, scuttles beneath the floor between the joists and occupies the cellar and the sewers. The

Starlings roost in Trafalgar Square, London

brown rat has a much wider taste in food, eating not only the vegetable matter favoured by the black, but meat as well. Today, it is dominant in most parts of most cities and the black rat has largely retreated to the docks, where its numbers are regularly reinforced by fresh immigrations from the colonies still flourishing on sea-going ships.

Successful though rats and pigeons, termites and geckos have been, the number of animal species that have solved the problems of urban living is small compared with the huge numbers of species that live in any one of the natural environments. The supply of food in towns, however, is lavish and continuous throughout the year. As a result, those species that do live in towns often multiply prodigiously. So cities are frequently overrun by plagues. Rats, protected within buildings from seasonal variations in the weather, breed all the year round, producing litters of up to twelve babies every eight weeks or so. Pigeons, even though they live outside, nonetheless manage to lay eggs several times each year and may nest in any month, winter or summer.

The endless proliferation of such creatures brings great problems to those who built the cities for their own habitation. Rats and mice raid the food stores and contaminate even more than they eat. Pigeon droppings corrode stone and masonry and disfigure and damage buildings. But there are even graver problems. Since neither rats nor pigeons in towns have any major predators, individuals that are crippled or handicapped by disease are not quickly killed and eaten, but survive for a long time, spreading their infections. So with the plagues comes disease. Rats carry fleas which not only bite them but also bite human beings. In the fourteenth century such fleas transferred the bubonic plague from rats to men, and one quarter of the entire population of Europe died as a consequence. Less than a century ago, a similar rat-carried disease killed eleven million people in India. Pigeons, though they have not been responsible for such horrifying epidemics, are nonetheless also disease carriers, and suffer from paratyphoid as well as the pigeon pox that produces crippling growths on their feet. The tribes of mangy wild dogs, descendants of domesticated creatures, that roam the streets of many cities, can carry that most dreaded of diseases, rabies. Urban man, for the sake of his own survival, has no alternative but to control these animal populations in his cities.

Not many people object to the eradication of clothes moths or deathwatch beetles. Few believe that it is morally wrong to kill rats and mice when they invade our homes and steal from our larders. More people are upset when pigeons are netted and killed, even though they may be reckoned almost as damaging and dangerous as rats. Nonetheless, most of us now accept that we have to manage the balance of animal populations in our cities and realise that this may, on occasion, require culling them.

But, happily, management can also involve encouraging others to live. We want a greater variety of creatures around us in our artificial world, so we set aside parks, plant trees, put up nest boxes, plant special flowers to entice butterflies and manage our gardens in such a way that they support the wild creatures in which we are interested. Indeed, the authorities in many cities have recognised their responsibilities as the controllers of the varied non-human populations within their boundaries.

But the countryside is also our creation. That too has to be managed. For centuries, decisions about what should live there were made separately by many different people, seldom in co-ordination with others or with any clear idea of the long-term effects of their actions. Only now, very late in the day, are we trying to put together a national policy which will take account of the advice of biologists who have some knowledge of the dynamics and inter-relationships of animal and plant populations, and which will have regard for the interests of all who use the land.

Yet even these large-scale decisions, to be truly effective, cannot be taken in national isolation. One country may protect the breeding grounds of a migratory bird very efficiently but the birds may still be exterminated if another country allows them to be shot when they reach their winter feeding grounds. Lakes will not remain full of fish even when those living on their shores take the greatest trouble to prevent pollution, if factories in another country spout their fumes so high into the air they contaminate the clouds and turn the rain, which falls days later and hundreds of miles away, into acid.

And still, even when these chains of cause and effect are acknowledged, the belief persists that beyond the towns, beyond the tamed countryside of the developed world, the natural world is so vast that it can survive any pillage, so resilient that it can recover from any damage. How false that is has been demonstrated again and again.

Some of the most fertile waters in the world lie just off the coast of Peru, around two groups of islands, the Chinchas and the Sangallans. Here an ocean current brings up nutrients from the deep sea floor to the surface in much the same way as happens on the Grand Banks of Newfoundland and with much the same result. Plankton blooms and supports great shoals of fish. The major direct consumers of the plankton are small shoaling fish, the anchovetas. These, in turn, are eaten by bigger fish such as sea bass and tunny and by vast numbers of birds which roost and nest on the bare rocks of the islands. Terns, gulls, pelicans and boobies swarm in huge flocks. Most numerous of all, fifty years ago, was a kind of cormorant called the guanay. Five and a half million of these birds alone nested there. Unlike the gannets and the pelicans, the guanay does not range far for its food, nor does it dive deep for it. It gets all it requires from the shoals of anchovetas, swimming close by and near the surface.

The guanay's digestion is odd and, it seems, not very efficient, for it absorbs only a relatively small proportion of the nutriment in the anchovetas it catches and excretes the remainder. The greater part of its droppings fall into the sea, where they fertilise the water and promote still further the growth of the plankton. But about a fifth of the guanay's droppings fall on to the rocks of the islands. Rain rarely falls in this part of Peru. In consequence, the droppings do not wash away but accumulate, forming deposits that were once over 50 metres deep. The Indians on the mainland, in pre-Columbian times, knew very well that this was a magnificent fertiliser and used it on their plantations. It was not until the nineteenth century that other peoples made the same discovery. Guano, as it was called, proved to be thirty times richer, in terms of nitrogen, than ordinary farmyard manure and contained many other important elements besides. It was exported all over the world. Distant countries based whole agricultural industries on it. Its price rose and rose. Sales of guano abroad contributed more than half of Peru's

national income. And fleets of fishing boats working around the islands harvested the sea bass and the tunny to provide food for people all over Peru. It would have been difficult to find a richer, more productive natural treasury anywhere.

Then, some thirty years ago, chemical fertilisers were developed and marketed. Though not as good as guano, nonetheless the price of guano began to fall and some people on the coast decided it would be marginally more profitable to harvest the anchovetas instead. They were not suitable for human consumption, but they could be turned into meal which would be eaten by chicken, cattle and pet animals. Netting the gigantic shoals was only too easy. The fishing was uncontrolled. In a single year, 14 million tons of anchovetas were hauled out of the waters. Within a few years, the shoals had all but disappeared. The guanays, in consequence, starved. Millions of the birds were washed up dead along the Peruvian coast. The survivors were so few in number that they no longer produced enough guano to make its collection worthwhile and the guano market collapsed completely. Neither were there enough guanay birds to fertilise the sea and sustain the plankton at its previous levels, so even though the anchovy fleet has stopped fishing, the recovery of the shoals is by no means assured. Certainly it will not be swift. Mankind, by not accepting his responsibility to manage, has succeeded in damaging not only the guanay, the anchoveta and the tuna, but himself.

The other great natural resource of the world, second only to the oceans, is the tropical rain forest. That too is being plundered in a similarly reckless way. We know that it plays a key role in the worldwide balance of life, absorbing the heavy equatorial rains and releasing them in a steady flow down the rivers to irrigate the lower fertile valleys. It has given us immense riches. Some 40 per cent of all the drugs we use contain natural ingredients, many of them deriving from the forest. Timber from the trunks of its trees is the most valued of all wood. For centuries, foresters have collected it, seeking particular kinds of trees, pulling them out and leaving the rest of the forest community little damaged. They planned their activities carefully so that they did not return to the same area for several years and gave the forest time to recover.

But now pressures on the rain forest have intensified. The increase of human beings in the surrounding countryside has led, understandably, to more and more of the jungle being cut down so that the land can be used to grow food. As we now know, the fertility of the jungle lies more in the substance of its plants than in its leached-out soil and the cleared land becomes exhausted and infertile after a few years. So the people fell more forest. Adding to this encroachment, modern machinery makes it easier than ever before to turn timber into cash. A tree that took two centuries to grow can now be knocked down in an hour. Powerful tractors can drag the fallen trunks out through dense forest with comparative ease, even if, in the process, they destroy many other trees that have no immediate cash value. So the jungle is disappearing at a swifter rate than ever before. Every year an area the size of Switzerland is being cut down. Once it has gone, the roots of the trees no longer hold the soil together. The lashing rains wash it away. So the rivers turn to brown roaring torrents, the land becomes a soil-less waste and the richest treasury of plants and animals in the world has vanished.

Guanay cormorants, Peru

The roll call of such ecological disasters could be extended almost endlessly. It is only too easy to demonstrate the damage we have now inflicted on the wildernesses of the world. It is more important to consider what should be done about it.

We have to recognise that the old vision of a world in which human beings played a relatively minor part is done and finished. The notion that an ever-bountiful nature, lying beyond man's habitations and influence, will always supply his wants, no matter how much he takes from it or how he maltreats it, is false. We can no longer rely on providence to maintain the delicate interconnected communities of animals and plants on which we depend. Our success in controlling our environment, that we first achieved 10,000 years ago in the Middle East, has now reached its culmination. We now, whether we want to or not, materially influence every part of the globe.

The natural world is not static, nor has it ever been. Forests have turned into grassland, savannahs have become deserts, estuaries have silted up and become marshes, ice caps have advanced and retreated. Rapid though these changes have been, seen in the perspective of geological history, animals and plants have been able to respond to them and so maintain a continuity of fertility almost everywhere. But man is now imposing such swift changes that organisms seldom have time to adapt to them. And the scale of our changes is now gigantic. We are so skilled in our engineering, so inventive with chemicals, that we can, in a few months, transform not merely a stretch of a stream or a corner of a wood, but a whole river system, an entire forest.

If we are to manage the world sensibly and effectively we have to decide what our management objectives are. Three international organisations, the International Union for the Conservation of Nature, the United Nations Environmental Programme, and the World Wildlife Fund, have cooperated to do so. They have stated three basic principles that should guide us.

First, we must not exploit natural stocks of animals and plants so intensively that they are unable to renew themselves, and ultimately disappear. This seems such obvious sense that it is hardly worth stating. Yet the anchoveta shoals were fished out in Peru, the herring has been driven away from its old breeding grounds in European waters, and many kinds of whales are still being hunted and are still in real danger of extermination.

Second, we must not so grossly change the face of the earth that we interfere with the basic processes that sustain life – the oxygen content of the atmosphere, the fertility of the seas – and that could happen if we continue destroying the earth's green cover of forests and if we continue using the oceans as a dumping ground for our poisons.

And thirdly, we must do our utmost to maintain the diversity of the earth's animals and plants. It is not just that we depend on many of them for our food – though that is the case. It is not just that we still know so little about them or the practical value they might have for us in the future – though that, too, is so. It is, surely, that we have no moral right to exterminate for ever the creatures with which we share this earth.

As far as we can tell, our planet is the only place in all the black immensities of the universe where life exists. We are alone in space. And the continued existence of life now rests in our hands.

The earth

GREENLAND
ARCTIC CIRCLE
ALASKA
Iceland
CANADA
Mt St Helens
Grand Banks
UNITED STATES of AMERICA
The Rockies
Hudson
Mississippi
Mojave Desert
Sonoran Desert
Azores
MID-ATLANTIC RIDGE
ATLANTIC
TROPIC OF CANCER
Mexico
GULF OF MEXICO
Hawaii
Yucatan Penisular
Cuba
West Indies
PACIFIC
CARIBBEAN SEA
OCEAN
Mopti
Isthmus of Panama
Orinoco
150°
EQUATOR
120°
90°
60°
30°
Galápagos Islands
Amazon
BRAZIL
Ascension
OCEAN
PERU
St Hel
Tahiti
The Andes
Iguazu Falls
Atacama Desert
TROPIC OF CAPRICORN
Easter
Plate
ATLANTI
ARGENTINA
Tristan da C
Gough
OCEAN
Patagonia
South Sandwich Islands
South Shetland Islands
ANTARCTIC CIRCLE
AN
CHAPTER 1
mountain chains
volcanoes
CHAPTER 2
snow and ice
tundra
CHAPTER 3
subpolar forest
temperate forest
CHAPTER 4
tropica rain forest

APTER 5 grassland

CHAPTER 6 desert

CHAPTER 8 fresh water

CHAPTER 11 salt water

ACKNOWLEDGEMENTS

My debts incurred in writing the preceding pages are many and large. Most importantly they are to my colleagues in BBC Television with whom the initial scripts were debated. On many occasions, they suggested new and unfamiliar creatures to replace the better-known examples that I first used, and pointed out gaps and misapprehensions in my first drafts. Their names are printed opposite. My gratitude goes particularly to those who took primary responsibility for individual programmes – Richard Brock, Ned Kelly and Andrew Neal – but, one way or another, I have debts to them all and I thank them all.

Both they and I, of course, are ultimately indebted to the innumerable scientists who have laboured over lifetimes to piece together coherent descriptions of animal communities in different environments and painstakingly elucidated the way in which they function. For the most part we have learned of their discoveries from their writings in specialist journals, but in some fortunate instances, we were lucky enough to work with such researchers in the field. On every occasion that that has happened we were met with the most generous and uninhibited help for which we are all deeply grateful. I personally have special gratitude to: Dr Jim Stevenson in Aldabra; Dr Nigel Bonner and Peter Prince in Antarctica; Dr Norman Duke in Australia; Dr Francis Howarth in Hawaii; Dr Putra Sastrawan in Indonesia; Truman Young in Kenya; Dr Mary Seely in Namibia; Dick Veitch in New Zealand; Dr Felipe Benevides in Peru; and Gary Alt, Prof. John Edwards, Prof. Charles Lowe and Prof. Robert Paine in the United States of America.

Dr Robert Attenborough, Dr Humphrey Greenwood, Gren Lucas and Dr L. Harrison Matthews have been kind enough to read individual chapters and steer me away from error. Crispin Fisher from Collins, and Stephen Davies and Susan Kennedy from BBC Publications, added not only accuracy but intelligibility to the text; and Jennifer Fry and Veronica Loveless discovered and displayed the illustrations with assiduity and flair. I am very grateful indeed to them all.

The Living Planet
Production Team

Executive Producer
Richard Brock

Producers
Ned Kelly
Andrew Neal

Assistant Producers
Ian Calvert
Richard Matthews
Adrian Warren

Production Team
Diana Richards
Marney Shears
Beth Huntley
Nicola Holford

Cameramen
Martin Saunders
Hugh Maynard

Assistant Cameraman
Jeremy Gould

Sound Recordists
Lyndon Bird
Keith Rodgerson

Film Editing
Andrew Naylor
David Barrett
Susanne Outlaw
Peter Simpson
Nigel Kinnings

Dubbing Mixer
David Old

Music
Elizabeth Parker
BBC Radiophonic Workshop

Graphic Designer
Margaret Perry

Unit Manager
Andrew Buchanan

Main Specialist Cameramen
Wolfgang Bayer
Stephen Bolwell
Rodney Borland
Robert Brown
Hugh Miles
Neil Rettig

Picture credits

Picture research: Jennifer Fry

Page 8 Picturepoint; 10 Jacana; 12 John Cleare, Mountain Camera; 14 David Attenborough; 16–17 NASA; 19 Jacana; 22–23 the late Prof. S. Thorarinsson; 25 A. C. Waltham; 27 Bruce Coleman (Fritz Vollmar); 28 Steve Terrill; 31 Seaphot/Planet Earth (Robert Hessler); 32 Bruce Coleman (J. Mackinnon); 35 Bruce Coleman (Keith Gunnar); 37 John Marshall; 41 Oxford Scientific Films (Doug Allan); 43 Seaphot/Planet Earth (Jonathan Scott); 45 Aquila (Graham Lenton); 47 Bruce Coleman (Gunter Ziesler); 49 Bruce Coleman (Francisco Erize); 51 Survival/ Anglia (C. Buxton/A. Price); 53 Ardea (Edwin Mickleburgh); 57 Eric Hosking; 58 Nature Photographers (M. E. J. Gore); 60 Robert Harding Associates (Wally Herbert); 62 Frank Lane (McCutcheon); 64 Ardea (Martin W. Grosnick); 66 Hannu Hautala; 69 Ardea (S. Roberts); 71 Ardea (J-P. Ferrero); 73 Ardea (C. R. Knights); 74 Bruce Coleman (Pekka Helo); 79 Bruce Coleman (Wayne Lankinen); 81 Aquila (Wayne Lankinen); 86 Tony Morrison; 89 Biofotos (Heather Angel); 91 Bruce Coleman (Gunter Ziesler); 93 Oxford Scientific Films (Michael Fogden) 95 Bruce Coleman (Gunter Ziesler); 96 Adrian Warren; 99 *top* Animals Animals/OSF (Stouffer Productions), *bottom* Bruce Coleman (John Mackinnon); 101 Oxford Scientific Films (Michael Fogden); 102 Tony Morrison; 104 Mantis Wildlife Films; 106 Ardea (François Gohier); 108 Bruce Coleman (Alan Root); 110 Tony Morrison; 112 Bruce Coleman (Leonard Lee Rue III); 115 Bruce Coleman (R. K. Murton); 117 Jacana; 118 Ardea (François Gohier); 121 Jacana; 122 Adrian Warren; 125 Ardea (Adrian Warren); 127 Wolfgang Bayer; 130 Bruce Coleman (Des Bartlett); 132 Biofotos (J. M. Pearson); 134–135 Seaphot/Planet Earth (Jonathan Scott); 138 NASA; 140 David Attenborough; 142–143 Andrew Neal; 145 Jacana; 147 Ardea (Wardene Weisser); 149 Bruce Coleman (Jen & Des Bartlett); 151 Andrew Neal; 153 Bruce Coleman (Jen & Des Bartlett); 155 Bruce Coleman (David Hughes); 157 Bruce Coleman (Carol Hughes); 158 *both* Rodger Jackman; 162–163 Natural Science Photos (C. Banks); 165 Aspect Picture Library (Carmichael); 169 NHPA (Stephen Dalton); 171 Bruce Coleman (Mik Dakin); 172 Press-tige Pictures (Tony Tilford); 174–175 British Antarctic Survey (J. P. Croxall); 177 Animals Animals/OSF (C. C. Lockwood); 179 Bruce Coleman (WWF/Y. J. Rey-Millet); 181 Bruce Coleman (R. Tidman); 182 Survival Anglia/OSF (Ashish Chandola); 184 Bruce Coleman (Jeff Foott); 187 David Attenborough; 188 Space Frontiers; 190 Bruce Coleman (Kim Taylor); 193 Bruce Coleman (Francisco Erize); 195 Oxford Scientific Films (Gerald Thompson); 197 Ardea (Adrian Warren); 199 David Attenborough; 201 A. van den Nieuwenhuizen; 203 Wolfgang Tins; 205 Oxford Scientific Films (J. A. L. Cooke); 207 ZEFA (Rosenfeld); 209 Alan Hutchinson Library (Jesco vom Puttkamer); 211 Bruce Coleman (H. Rivarola); 213 Frank Lane (E. Eisenbeiss); 214 Oxford Scientific Films (Avril Ramage); 218 Nature Photographers (W. S. Paton); 221 Biofotos (Heather Angel); 223 Ardea (McDougal); 225 Mantis Wildlife Films; 227 Seaphot/Planet Earth (Keith Scholey); 231 Photographic Library of Australia; 233 Anne Wertheim; 234 Oxford Scientific Films (G. I. Bernard); 238 Biofotos (Heather Angel); 240–241 Bruce Coleman (David Hughes); 244–245 David Attenborough; 247 Bruce Coleman (Christian Zuber); 249 Ardea (Liz & Tony Bomford); 251 Jacana; 252, 254–255 David Attenborough; 258 Bruce Coleman (Jennifer Fry); 260 R. T. Shallenberger; 262 Biofotos (Heather Angel); 264 Bruce Coleman (M. F. Soper); 268 Oxford Scientific Films (Peter Parks); 271 Biofotos (Heather Angel); 272 Seaphot/Planet Earth (Peter Scoones); 274 Seaphot/Planet Earth (Herwarth Voigtmann); 276 Animals Animals/OSF (Z. Leszczynski); 278 Bruce Coleman (Jane Burton); 280–281 Oxford Scientific Films (Laurence Gould); 283 Seaphot/Planet Earth (Herwarth Voigtmann); 285 Seaphot/Planet Earth (Peter David); 287 Dept. of Fisheries and Oceans, Canada; 290 Jacana (Claude Pissavini); 293, 296 Andrew Neal; 299 Susan Griggs Agency (Dimitri Ilic); 301 Biofotos (Heather Angel); 302 Survival Anglia/OSF (M. Wilding); 306 Oxford Scientific Films (Ronald Templeton); 309 Associated Press.

INDEX

The scientific name of an organism is shown in brackets in those cases where it differs considerably from the common name and where it has not been given in the main text. **Bold** figures refer to illustrations